GENIUS ENGINEER

THE STEP-BY-STEP GUIDE TO TRANSFORM

YOU INTO A

GENIUS ENGINEER

ACE YOUR PROJECTS, PUBLISH, PATENT & STARTUP

with real
life stories
of Great
Inventors

Rajaram Thiruvengadam, PhD.

ISBN: 9781093911077
   Imprint: Independently published

About the Author: Dr. Rajaram Thiruvengadam has over 20 years experience in teaching and research connected to Environmental Impact Assessment and Environmental Sustainability. He is currently Dean of Research at Roever Engineering College, Roever Group of Institutions, India. He has published several impactful papers from his research in the area of Sustainability Linkages. He earned his Bachelor of Civil Engineering with a Gold Medal and his Master of Environmental Engineering while topping his class, both from National Institute of Technology, Surat, India. His detailed profile can be accessed at:
http://tinyurl.com/m2t4y93
https://www.linkedin.com/in/rajaramenviroprof/
https://www.researchgate.net/profile/Rajaram_T
Comments and suggestions are welcome at rajaenviro@gmail.com

Dedication

To
Mother Earth

(Ala, Asase, Amalur, Atabey, Asherah, Bhoomithaai, Bhudevi, Eingana, Gaia, Izanami, Joro, Otukan, Pachamama, Phrae Mae Thorani, Wakan Tanka,.............)

For the patience, you are showing against the assault by your favorite children, *Homo sapiens*.

CONTENTS

Preface

Genius: extraordinary intellectual power especially as manifested in creative activity

Engineer: to contrive or plan out usually with more or less subtle skill and craft (Merriam-Webster)

Engineering as a profession is undergoing vast changes as progress in made in technology. However, the basic skill necessary to succeed as an engineer is the same old ingenuity or creativity. Nevertheless, the students of engineering are faced with the enormous task of first assimilating all the vast knowledge generated over centuries. Hence taking time to practice or cultivate the creative aspect of actually engineering something is often relegated to the final year of study. By which time it is often late to indulge in any imaginative or creative endeavor due to the pressure of meeting the project submission deadlines.

The idea to write this book arose from my struggles as an engineering student trying to figure out the purpose of the whole curriculum, when we were fed with seemingly unrelated wide-ranging courses. By the time I understood the true purpose of engineering, was out of college. Later as a teacher and project supervisor/ advisor to many batches/ individual students, felt that the limited interactions were not enough to convey the entire idea of 'how to enjoy the creative process of an engineering project'. Add to this the disconnect that exists between the students and faculty, due to differences in expectations. The faculty advisor is always pressed for time and students expect clear concise guidance for an activity that is supposed to be a journey of self-discovery and accomplishment. This book is not meant to be a substitute for a faculty advisor. Rather it is meant to equip the student with the basics of engineering a project, so that they can indulge in much more higher-level discussions with their faculty advisors.

Hence, this book is intended to open the eyes and minds of engineering students to the wonderful universe of engineering a solution. This book has been written to aid all the students across the ability/ aptitude spectrum. Therefore, some may find the steps in this book of very preliminary in nature and some may find it very exhaustive to the point of being intimidating. However, the 50 steps have been conceived to enable all types of learners to benefit from the joy of experimenting and enriching their experience of becoming an engineer.

So look beyond the goals of achieving good grades or landing a job and just indulge in enriching your knowledge, give wings to your creative dreams, shed your inhibitions and fear of being judged or failing at something. As they say, if you strive and achieve excellence in what you do, money and fame will chase you. Just Engineer!

WATCH OUT FOR:

Inspiring Inventors / Inventions: Real life inspiring stories & interesting facts about Inventors and their Inventions

Inventive Hominid: Timeline of historic inventions made by our human species – Homo sapiens. As they say, "it's in our genes."

Search alert: search the topic to get the latest information and do not forget to suffix the term with the current year

Note down: write your thoughts about the information asked for follow up and involved understanding

STEP -1

What is Engineering?

The term engineering is derived from the Latin Ingenium, meaning "cleverness" and ingeniare, meaning, "to contrive, devise". -IAENG, International Association of Engineers.

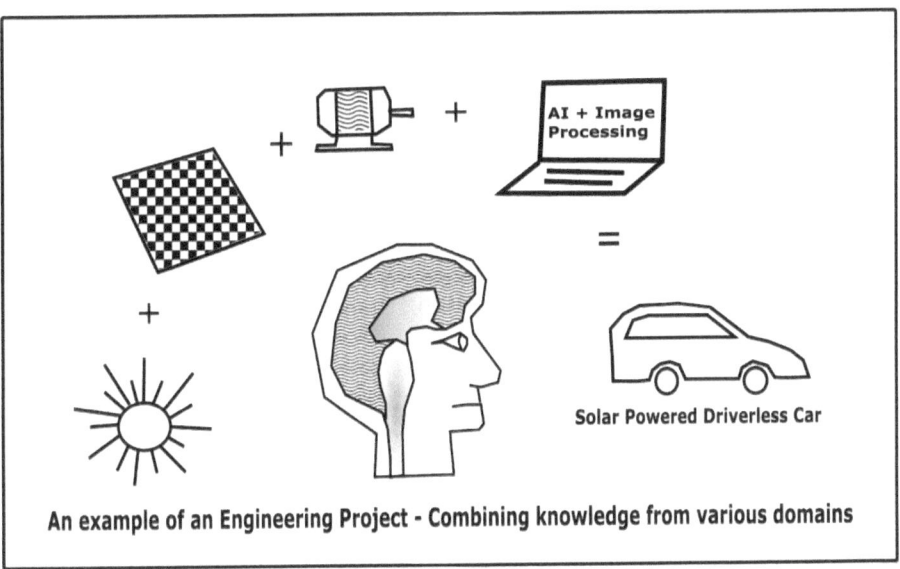

An example of an Engineering Project - Combining knowledge from various domains

Learning Objectives

After studying the contents of this chapter, you should be able to:

- Understand the concept of Engineering
- Understand the difference between science and engineering
- Appreciate the significance of an engineering project
- List out the attributes an engineering graduate should possess
- Understand the importance of starting the project early in the program

Engineering is The creative application of scientific principles to design or develop structures, machines, apparatus, or manufacturing processes, or works utilizing them singly or in combination; or to construct or operate the same with full cognizance of their design; or to forecast their behavior under specific operating conditions; all as respects an intended function, economics of operation and safety to life and property (EPCD, 1947, -1.1).

Phew! That must have given your head a real spin. Not to worry, engineering is all about providing a solution to a problem. Simple, is it not?

Engineers need not worry about the fundamentals of scientific theories; scientists are there worrying about it. Engineers should rather utilize the available scientific knowledge to provide real world solutions.

Finding out the thermal properties of microwave is science, but turning that knowledge into a microwave oven is engineering.

Studying the properties of steel under tensile stress is science, but using that knowledge to design magnificent steel structures like the Eiffel Tower is engineering.

We can go on and on with examples. Therefore, to be a successful engineer, you should demonstrate your problem solving skill through your project.

What is a project?

As per the Project Management Institute (pmi.org), the world's leading Project Management Certification Organization [-1.2],

"It's a temporary endeavor undertaken to create a unique product, service, or result."

Why Engineering Project is Important

Engineering project that you take on as a compulsory course is the one that demonstrates your ability to combine the concepts that you learnt in various courses to engineer a solution to a problem.

Hence an important indicator of your ability to engineer solutions in your chosen area of specialization is how good and innovative is the project you have accomplished.

Now in most of the universities, majority of the engineering courses are designed with some weight-age for an individual or a group project. This is done to stimulate the thinking of the scholars into engineering mode. So that, by the time they get started with their major project in the final year, they are aware of its significance and the methodology to accomplish it.

What should you demonstrate through your project?

An engineering graduate is required to demonstrate specific skills which are defined by Accreditation Board for Engineering and Technology (ABET). The Engineering project or capstone project is a crucial opportunity through which these skills can be effectively demonstrated. These skills defined as student outcomes by ABET [-1.3] are:

- An ability to identify, formulate, and solve complex engineering problems by applying principles of engineering, science, and mathematics
- An ability to apply engineering design to produce solutions that meet specified needs with consideration of public health, safety, and welfare, as well as global, cultural, social, environmental, and economic factors
- An ability to communicate effectively with a range of audiences
- An ability to recognize ethical and professional responsibilities in engineering situations and make informed judgments, which must consider the impact of engineering solutions in global, economic, environmental, and societal contexts
- An ability to function effectively on a team whose members together provide leadership, create a collaborative and inclusive environment, establish goals, plan tasks, and meet objectives
- An ability to develop and conduct appropriate experimentation, analyze and interpret data, and use engineering judgment to draw conclusions
- An ability to acquire and apply new knowledge as needed, using appropriate learning strategies.

Do not wait until your final year to start your project

In most of the universities, the project requirement is mostly in the final year, when they are formed into groups and assigned a project advisor. Nevertheless, why wait until the final year, when the time might not be enough to go through a number of iterations and accomplish something big. Hence, start thinking about your project from the time you enter an engineering course. Though you might not have any faculty advisor from first year, this book is meant to give you step-by-step guidance to dream and execute your project. So, shall we jump right in?

Inspiring Inventions

History of Engineering Discipline

Engineering initially developed for military purposes such as building forts, defense structures, and contraptions. Later Civil Engineering branched out to deal with structures for civil purposes, such as bridges, roads, etc, [-1.4].

As technology developed the branches of mechanical and electrical engineering were created.

Further explosion of technology has given rise to branches like mining, chemical, automotive, petroleum, marine, manufacturing, electronics, aeronautical, nuclear, textile, communication, information, biotech, bio-medical, environmental, software, robotics, etc.

Though the terms engineering and technology are used interchangeably, technology concerns with applying practical knowledge in a particular area. Whereas, Engineering is about solving a wide range of problems by applying scientific knowledge.

STEP 0

Let us begin: Stages of a Project

A conceptual model is neither idle nor faithful: it is, or rather it is supposed to be and So taken until further notice, an approximate representation of a real thing -Mario Bunge (1972) Philosophy of Physics

Learning Objectives

After studying the contents of this chapter, you should be able to:
- Understand the stages of an engineering project
- Understand the types of projects
- List out the stages of a theoretical and patentable project
- Know the chapters connected with the stages of a project

In this chapter, we shall understand the sequence of activities you have to follow in executing your engineering project. The sequence is presented in the form of a flow chart and the activities are indicated along with the chapter number in which they are explained in detail. The project sequence is further explained in terms of two types of project: an investigative/ theoretical type that dwells deeper into the conceptual realm useful for those with an aptitude for research and higher studies, and a product/ technology oriented patentable/startup-oriented type which emphasizes on using available knowledge to come up with an innovative product or service. Go ahead, understand both types, and jump in as per your interest.

The stages of a typical Engineering Project

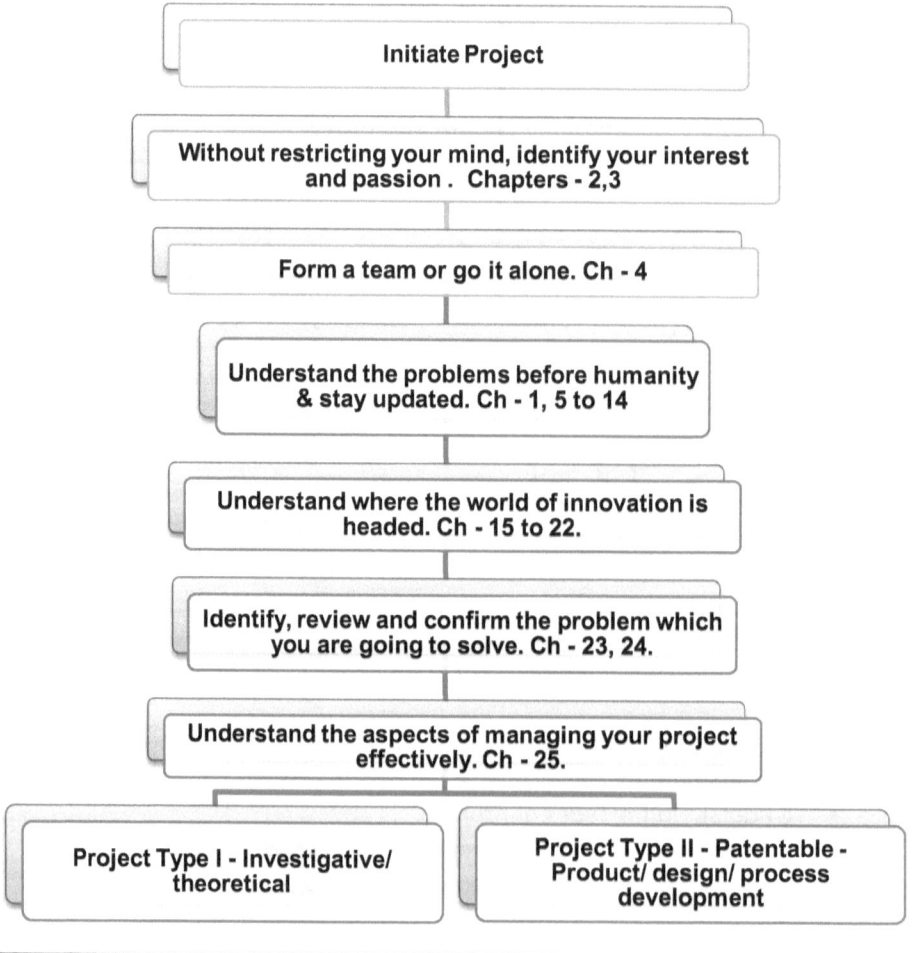

Initiate Project

Without restricting your mind, identify your interest and passion . Chapters - 2,3

Form a team or go it alone. Ch - 4

Understand the problems before humanity & stay updated. Ch - 1, 5 to 14

Understand where the world of innovation is headed. Ch - 15 to 22.

Identify, review and confirm the problem which you are going to solve. Ch - 23, 24.

Understand the aspects of managing your project effectively. Ch - 25.

Project Type I - Investigative/ theoretical

Project Type II - Patentable - Product/ design/ process development

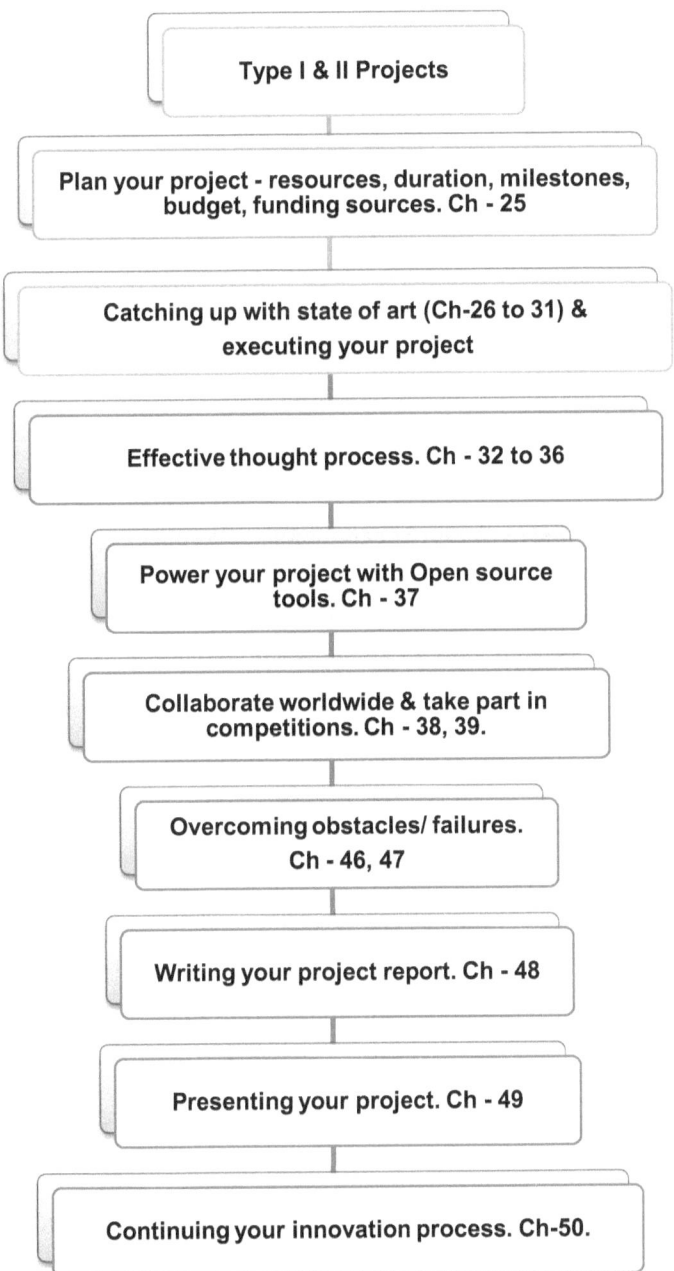

The stages of a Typical Engineering Project (continued)

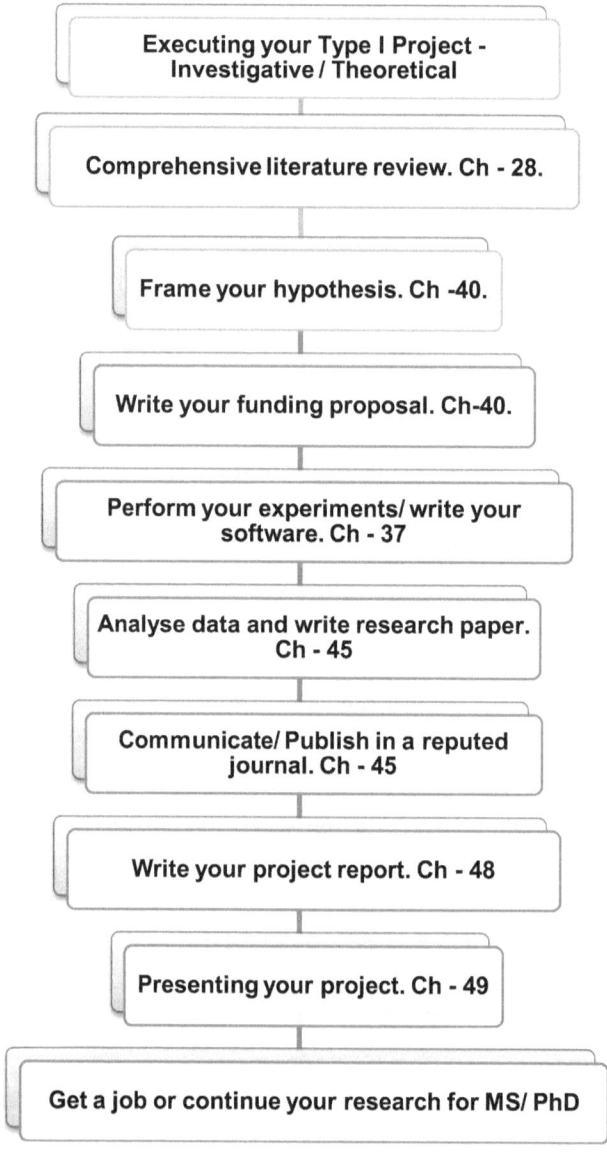

The stages of an Investigative/ Theoretical Project

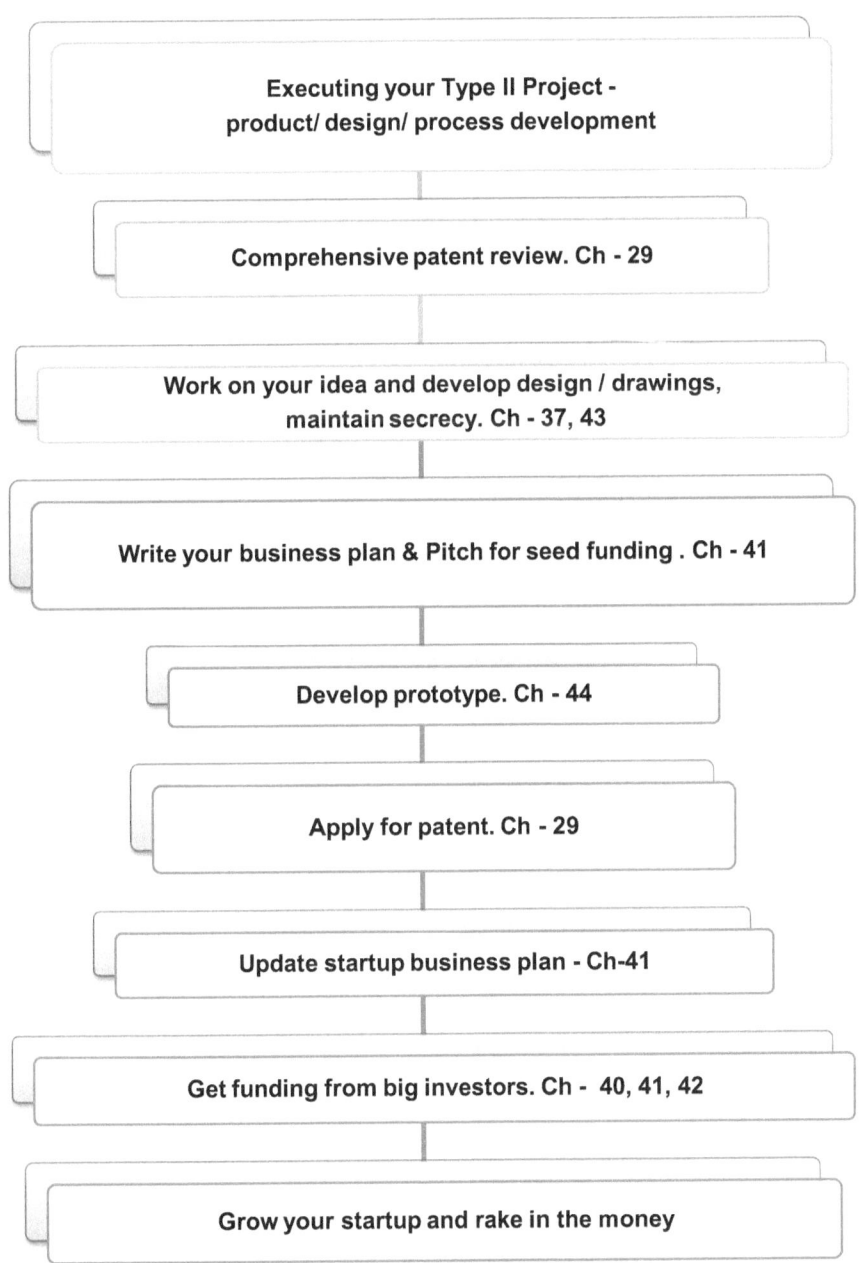

Executing your Type II Project -
product/ design/ process development

Comprehensive patent review. Ch - 29

Work on your idea and develop design / drawings,
maintain secrecy. Ch - 37, 43

Write your business plan & Pitch for seed funding . Ch - 41

Develop prototype. Ch - 44

Apply for patent. Ch - 29

Update startup business plan - Ch-41

Get funding from big investors. Ch - 40, 41, 42

Grow your startup and rake in the money

The stages of a patentable Project

STEP 1

Problem Solving: Engineering a Solution

A solution, which does not prepare for the next round with some increased insight, is hardly a solution at all.
-Richard Wesley Hamming (1997). The Art of Doing Science and Engineering

The Inventive Hominid (TIH) - First Stone tipped Spear (1 TR 1)

Learning Objectives

After studying the contents of this chapter, you should be able to:

- Understand the critical problems facing human society
- Appreciate the interconnectedness of many problems and think about providing an engineering solution to these problems

What is a problem?

"A matter or situation regarded as unwelcome or harmful and needing to be dealt with and overcome." -Oxford Dictionary

Now we are ready to solve the most pressing problems the world is facing. Attention everyone! Here we come at last to blow away your problems.

The great enduring problems of humankind

- **Poverty** – lack of money, food, water; wastage of food produce, over population, etc.
- **Pollution** – pollution of water, air, soil, oceans, etc.
- **Climate change** – release of CO2 from fossil fuel use, destruction of ecosystems, extinction of species
- **Diseases** – vector borne, waterborne, lifestyle diseases
- **Scarcity of water** – polluted water, saline water
- **Scarcity of energy** – lack of clean, affordable energy
- Asymmetrical availability of information, poor governance, corruption, offshore tax havens, etc.
- Specter of terrorism, cyber crimes, internet warfare

Relax! An engineer can earn his livelihood as long as there are plenty of problems to solve in this world. All these problems definitely have an engineering solution. You can choose any problem to contribute your solution. Should you only choose a problem from your field of study? Inspiring Inventors

Inspiring Inventors

The genius who lived and breathed Art & Innovation

Leonardo Da Vinci (1452-1519) is considered an Universal Genius whose areas of interest included invention, painting, sculpting, architecture, science, music, mathematics, engineering, literature, anatomy, geology, astronomy, botany, writing, history, and cartography[1.1]. He is considered a polymath genius who did not leave any field untouched with his investigations. Not content with his breathtaking paintings and sculpture, his pioneering research extended into anatomy, designs for aircrafts, armored warfare vehicles, musical instruments, mechanical soldier, hydraulic pumps, mortar shells, steam cannon, etc. His investigations are preserved in 13,000 pages of original notes and drawings, a part of which can be seen online as Codex Arundel [1.2].

STEP 2

Come Out Of Your Domain Mentality

"Engineering faculties should ensure that breadth of learning, beyond the technical aspects of the specialist engineering discipline, is a major thrust in engineering education",
Canadian Academy of Engineering (CAE) [2.1].

The Returning flight of the Boomerang depends upon the lift, spin, and precession. It is interplay of air pressure, velocity, angle of throw and shape of boomerang.

The Inventive Hominid

Boomerang, 50,000 BCE
Australian Aborigine Civilization

TIH - Invention and Use of Boomerang (2 TR 1, 2 TR 2)

Learning Objectives

After studying the contents of this chapter, you should be able to:

- Understand the interdisciplinary nature of world's problems
- Appreciate that most inventors did not specialize in any particular field, rather enhanced their knowledge based on the problems they were trying to solve
- Learn to think beyond your academic specialization

You can solve any problem, if it is in your passion zone!

A dangerous disease afflicting our engineering students and graduates alike is the disease of 'The Degree'. They restrict their thinking by the stream of engineering they have chosen to study. As pointed out earlier, engineering excellence needs a good conceptual grounding in the basic sciences – physics, chemistry, mathematics and you can include biology. Just because you were forced to choose or rather willingly, chose metallurgical or material engineering does not mean that problems outside your domain are a no go. Do not be limited by your syllabus when you choose your problem.

Would you like to end world poverty through technology? You are welcome! Get cracking to understand the problem and come up with a solution.

A 50 – 60 year career should not be limited to what you learned in 4 – 5 years. Have you not heard of the aeronautical engineer turned environmental engineer turned farmer? Such creatures do exist in the real world. Just browse the internet and you will find hundreds of engineers who have turned into farmers, fashion designers, writers, you name it and they have done it.

Engineering education is supposed to give us the ability to visualize, research and solve any problem. However, what about expert knowledge that is necessary to solve a problem? If your problem is from your passion zone, you can become an expert in weeks or months by absorbing the knowledge relevant in that area.

The great inventors of yore did not even get a degree from college; they just chose a problem and attacked it from all angles for months and years until they found a solution. So why should not we choose the problem which is interesting to us?

Once you choose the problem, you can always take courses necessary to enhance your knowledge, both within your university, outside and online. So, do not restrict your choice of a problem within your academic specialization, go ahead, dream big and realize your full potential.

Inspiring Inventors

A portrait painter who invented the single wire telegraph

Samuel Morse, the inventor of the single wire telegraph system and co-developer of the Morse code, was originally a successful portrait painter. At the age of 34, while he was painting a portrait in Washington DC, he received news of his wife being seriously ill and of her death the next day. His wife was buried before he could reach home. Heartbroken, Morse decided to explore a means of rapid long distance communication and ultimately came up with the single wire telegraph system [2.2].

STEP 3

How to Identify Your Passion Zone?

"The passion rebuilds the world for the youth. It makes all things alive and significant."- Ralph Waldo Emerson, "Love," Essays, First Series (1841).

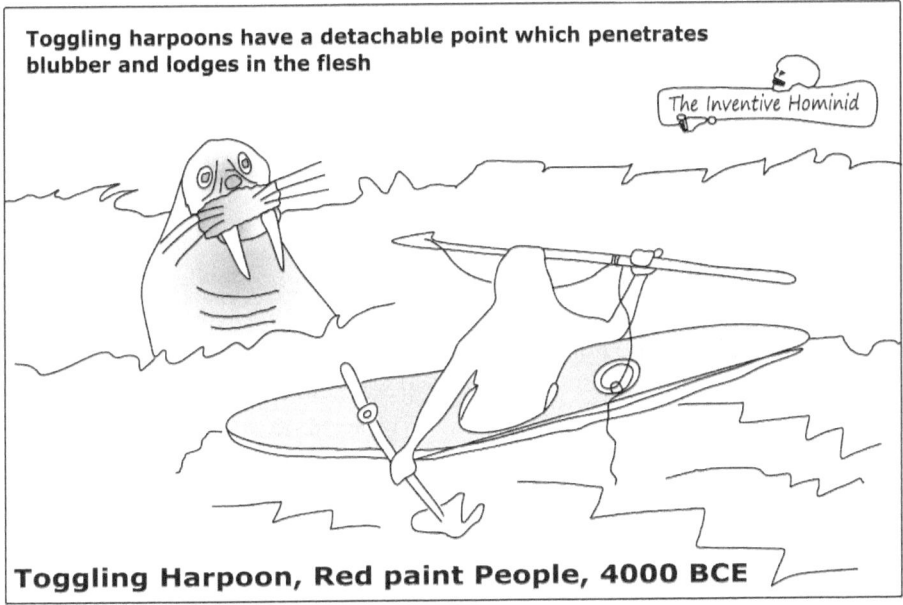

TIH - Toggling Harpoon (3 IR 1, 3 TR 1)

Learning Objectives

After studying the contents of this chapter, you should be able to:
- Identify your interest in the whole of engineering
- Find out that one area of knowledge you understand effortlessly and without much of a sweat

Passion zone? What is this?

Effortless understanding- go back to your school days, every one of us had a favorite subject/course that we just read through even before it was covered in class. We could just connect with the concepts without seeking help from anybody.

Acing tests without breaking a sweat – we used to score well in that course/ subject without hard work. Importantly, we used to ace questions, which were considered tough by our peers.

Hard to stop reading subjects – we used to just rush through the chapters in the book and be hungry for more or used to rummage through the subject with higher-class textbooks.

Subject feels like a long lost friend – now when you get an inkling of this favorite subject in engineering courses, you get excited and concentrate exclusively on delving into that chapter or unit.

Are you getting the hint: now identify the subject/ area, which ticks all the above. It could be physics or chemistry or biology or, more specifically semi-conductors or superconductors or nuclear chemistry or ecology or acoustics or probability or psychology or ergonomics or economics or Vedic history or English poetry, etc. and that subject is in your passion zone.

Have you identified your passion? Write it down.

I am passionate about:

Good, that is a wonderful subject, now start reading up its interface with engineering and definitely, you will find umpteen areas of the interface that your passion is having with your/ any area of engineering.

For example, if your passion is the study of languages, i.e. linguistics, you can straight away jump into Natural Language Processing (NLP), which deals about the interaction between computers and human languages. Computer science, artificial intelligence, natural language forms, etc. are the connected areas that contribute to NLP.

Inspiring Inventors

A boy who rose from poverty and lifted a nation into the stratosphere

Bharat Ratna Dr. A PJ Abdul Kalam (1931 – 2015) - an Indian scientist popularly known as People's President (2002 to 2007) was born and raised in Rameswaram, Tamil Nadu. Born in a family of limited means, Kalam supplemented his family income in his childhood by delivering newspapers around town [3.1]. Fascinated when his schoolteacher explained how a bird flies, he decided to study aeronautics. He studied physics and aerospace engineering in college. He dedicated himself to service of the nation through his life long tenure in DRDO (Defence research & development Organization) and ISRO (Indian Space Research Organization).

He was known as the *Missile Man of India* for his work on the development of ballistic missiles and Satellite launch vehicle technology [3.2]. He was also instrumental in developing low cost cardiac stent and Orthosis calipers from advanced materials. Dr Kalam was passionate about transforming rural India into sustainable regions and outlined his vision through the concept of PURA (Providing Urban Amenities to Rural Areas). He was fondly known as the People's President for his untiring efforts to reach out to the people and particularly students whom he addressed in the hundreds of thousands, often exhorting them to lead a principled life and contribute to the nation. A bachelor who never used his high offices for his personal gain, he led a Gandhian life of simplicity and promoted harmony among various religions by espousing secularism. He collapsed at age 83, due to cardiac arrest while addressing college students and passed away. His worldly possessions consisted of few clothes and thousands of books. His legacy lives on through the books he has written and the way he lived true to Gandhiji's saying, "Be the change you wish to see in the world" [3.3].

STEP 4

Form a Team: Should you work as a Group or go alone?

I believe in the essential unity of man and for that matter of all that lives.
-Mahatma Gandhi

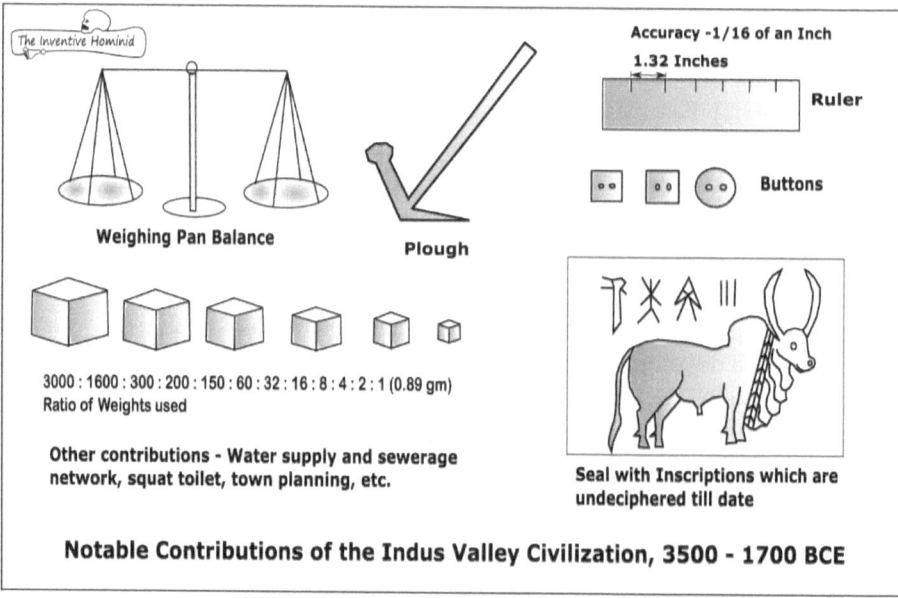

Notable Contributions of the Indus Valley Civilization, 3500 - 1700 BCE

TIH - Indus Valley Civilization (4 TR 1, 4 TR 2, 4 IR 1, 4 IR 2)

Learning Objectives

After studying the contents of this chapter, you should be able to:

- Appreciate the advantages and disadvantages of working as a team
- Understand the benefits and risks of working alone
- List the rules to enable a group to function effectively
- Understand aspects to ensure professionalism in your team

Why a team is better than going it alone

In a team, the energy levels are multiplied. Magnitude of team potential is more than the sum of the individuals

One member having down time does not affect progress, as other members will pick up the slack.

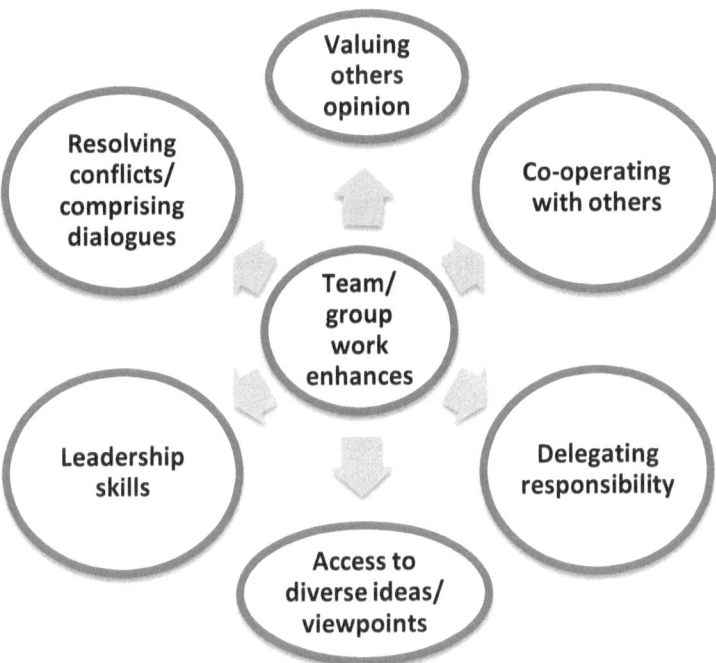

The team members might complement each other's skill sets. Every member brings a unique expertise not available with other members.

Working in a group or a team enables you to enhance your skills in these areas [4.1, 4.2]

- Valuing others opinions
- Delegating responsibility
- Solving conflicts among members
- Getting access to diverse ideas and creative thinking
- Finding and recruiting members with complementary skills

A group will enable the continued progress in a project even if one or two members are indisposed. It will enable higher achievements than is possible if members work individually. Hence, it is time that students look for likeminded people with skills that can strengthen all areas required for an innovative project.

Rules for effective group work
- Discuss and delegate responsibility
- Appoint each individual to lead in a specific area
- Respect every member's viewpoint
- Appreciate and give credit when it is due
- Ensure growth of all members

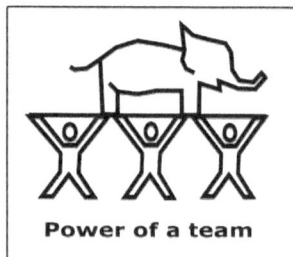

Power of a team

Get professional in working as a team

It is important to work and interact as professionals in a team environment. Some of the measures you need to adopt to enhance professionalism include:
- Start team meetings on time
- Take turns to lead the discussion at meetings
- Record the decisions taken by the team and ensure everybody signs it
- Agree responsibilities and deadlines for each member, record and circulate signed document
- Try to arrive at a consensus on important decisions
- Remember to disagree without being disagreeable
- The team comes first and everything next.

A typical format for recording decisions/ minutes of a meeting

Meeting of the Advanced Technology Team – Project Codename-YY99

Meeting No.　　　Date:　　Time:

Members Present: 1.　　2.　　3.　　4.

Sl. No.	Points discussed/ to be done	deadline	Member responsible
1			
2			
3			
4			
5			

Sign of members:
1.　　　　2.　　　　3.　　　　4.

Advantages of going it alone

"The right to be let alone is indeed the beginning of all freedom."-William O. Douglas

There are certain advantages of doing the project alone. These are:

- You are free to select a problem that interests you most
- You can work at your own pace according to your situation
- You do not have to wait for others to do their part
- You get all the credit for yourself

The disadvantages of working alone

- The project stalls if you are not able to work due to various reasons
- Like success, failure also is your own
- You cannot outperform a good synergistic team
- You will miss the company of others during difficult phases of the project
- Tight deadlines can be daunting to accomplish alone

Hence, consider all the options before deciding on a team or going it alone. Going alone might be suitable for those planning an academic or research career, where the next steps of acquiring MS and PhD are supposed to be individual projects. If you are planning on a startup, then working as a team will give you the required skills to lead a startup.

Inspiring Inventors

Kidnappings of scientists and inventors: Operation paperclip Vs Operation Osoaviakhim

After the fall of Nazi Germany in the World War II, there was a race between US and Russia to capture the maximum number of German Scientists, Engineers, and Technicians, take them to their country, and gain an advantage in developing advanced technologies.

Operation Paperclip [4.3] was a secret program of US Intelligence Agency carried out by its special agents. It recruited more than 1600 German Scientists, engineers and technicians that included Wenher von Braun and his V-2 Rocket team 2. They were taken to US and kept under government employment. They contributed greatly to the advancement of technology in a number of areas including the Saturn-V launch vehicle that enabled man to land on the moon.

Not to be outdone, Russia launched its own kidnapping operation in 1946 code named **Operation Osoaviakhim** [4.4]. It recruited under gunpoint, more than 2200 German experts numbering around 6000 including their family. The operation involved transporting the specialists along with their family, furniture, belongings, and test equipment in 92 trains. They contributed immensely in a number of technological areas including the most powerful turbo-prop engine ever built.

STEP 5

Motive for Problems Solving

The dependence of the individual upon society is a fact of nature that cannot be abolished - Albert Einstein, Why Socialism? (1949)

The Inventive Hominid

Adobe (mud brick), 3000BCE - South American Tribes

TIH - Adobe Mud Brick (5 TR 1, 5 TR 2)

Learning Objectives

After studying the contents of this chapter, you should be able to:
- Understand the various motives that enable people to solve problems
- Fix your motive for problem solving and list down the problem you would like to solve

Want a motive for problem solving?

Good, any Endeavour without a motive is doomed from the start. Motives depend on an individual and his personal values. However, whatever might be your values, you can identify with any of these three motives:

Societal Contribution

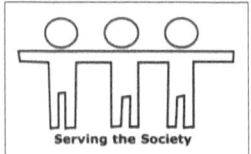
Serving the Society

Contributing your bit to our society is a great motive for many. People derive great satisfaction in abundance from the selfless contribution to society and this has been the single biggest motive for many of the greatest human beings who have graced our earth. Heard of Jonas Salk?

Fame

Achieving Fame

Who would not like to get some fame at the world level? Every human being aspires to be known and appreciated by fellow human beings. A yearning for appreciation is a fundamental necessity for a majority of our fellow humans to lead a happy and satisfying life. Alternatively, lack of appreciation is a major cause of depression in the lives of multitudes of people. Therefore, while you slog out there to solve a critical problem, the lure of sweet success and fame is a big motivator out there.

Money

Earning Money

Not interested in a societal contribution or fame? Yeah, we can understand. What is the use of toiling hard if at the end, we cannot smell the scent of money. It can solve any type of problem in every nook and cranny of our earth. How about coming up with an earth shattering invention, selling the patent for a billion dollars and buying an island in the south pacific to live happily ever after. Too good to be true! Is it not what we are after in our secret dreams? Now what are you waiting for, go for it.

Let us imagine that you have chosen a problem from the ones listed below:
- Ending world poverty through artificial intelligence in agriculture

- Solving the water crisis through robotic seawater desalinators
- Educating the millions of illiterate masses through creative content delivered to mobile phones
- Creating a sustainable local food network through artificial intelligence based on social media
- Wireless solar power transmission
- Virtual university for free education through distributed automated cloud based courses
- Remote, secure, and instant voting in elections and referendums
- Cancer cell removal through nano-bots
- Augmented reality based execution of engineering projects
- Enhanced space travel through teleporting
- Solving world hunger through artificial photosynthesis
- Knowledge and memory enhancement through embedded brain chips
- AI controlled drone army for sustainable pest control in agriculture
- Production, processing, and delivery of food through drones
- Artificial pollination through AI based insect-bots
- Write down the problems you want to solve.

How do we know as to what is a critical problem which needs urgent attention?

Inspiring Inventions

Men who eschewed profit to invent for the humankind

John Walker (1781-1859) was a British Inventor who invented the friction match. Though Walker trained to become a Surgeon, he left the profession and started studying chemistry. He invented the friction match consisting sticks coated with sulphur and tipped with a mixture of sulphide of antimony, chlorate of potash and gum. Walker refused to patent his invention and made it freely available to the public [5.1, 5.2].

Rev Patrick Bell (1799-1869) was a Church Minister and Inventor. In 1828, Bell successfully demonstrated his reaping machine pulled by two livestock and consisting of 12-vane reel, cutting knife and a canvas conveyor. Being a pious man, he never patented his invention and did not derive any financial benefit. However his design was patented in the USA and sold world over (5.3, 5.4].

STEP 6

Where to Look For a Problem

I am convinced that the United Nations provides the best road to the future for those who have confidence in our capacity to shape our own fate on this planet. -Former Secretary-General Kurt Waldheim

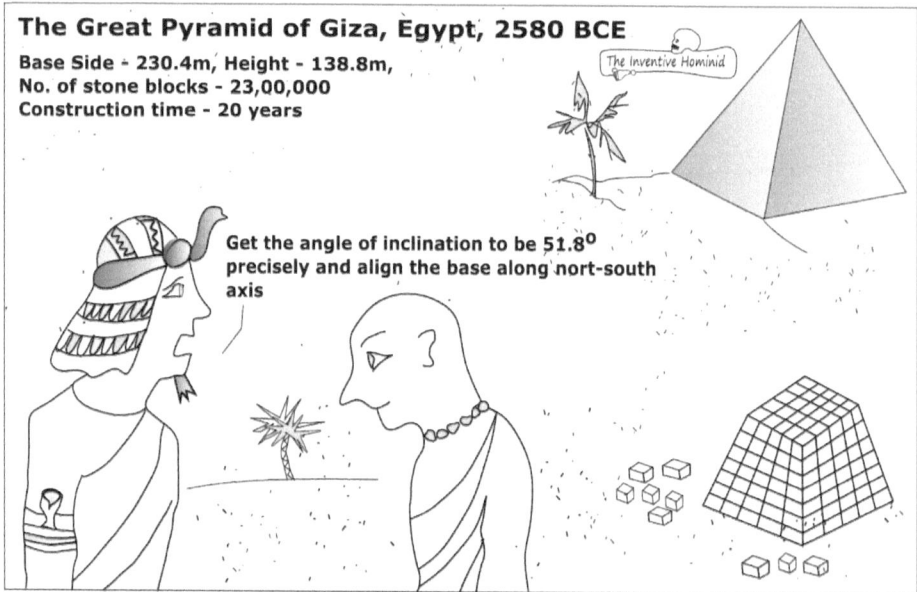

TIH - The Great Pyramid (6IR 1, 6TR 1)

Learning Objectives

After studying the contents of this chapter, you should be able to:
- Know the sources from where you can learn about current problems
- The role of news media, UN agencies, Think tanks, Government Departments in researching and releasing reports on major problems
- List out problems at local, regional, national, and global levels

Local, regional, national, international problems

News media

Newspapers are the best place to get a hang of local problems; they make a living by highlighting local conflicts.

Where there are conflicts, it is usually a problem involving a natural resource – land, water, minerals, food, forest, ocean, river, etc.

Just go through a couple of local newspapers to get a hang of the problems in that area.

Through newspapers on the internet, you can get a hang of problems in every corner of this globe from your desk. Want a search engine which will aggregate all news articles, try something like Google news, it does a good job of aggregating news articles from leading newspapers according to each country and language.

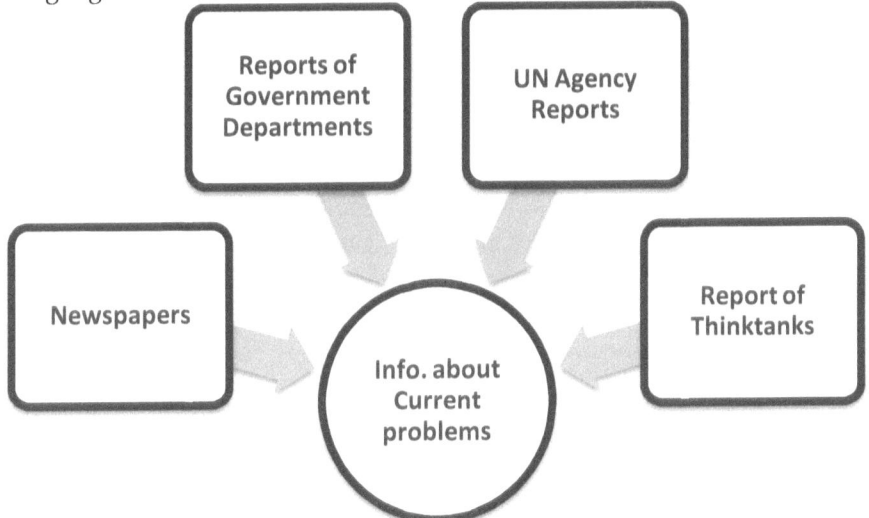

Try this website now: Google news

In the search box in Google news, type the name of your hometown and read all the news from your hometown.

Reports of UN Agencies, Think Tanks, Planning Commissions, etc.

Well, after getting a hang of local problems, now let us get the serious researched stuff from the experts. There is probably a UN agency for every important aspect of humanity.

- Food and Agriculture Organization (FAO)
- International Civil Aviation Organization (ICAO)
- International Fund for Agricultural Development (IFAD)
- International Labor Organization (ILO)
- International Maritime Organization (IMO)
- International Monetary Fund (IMF)
- International Telecommunication Union (ITU)
- United Nations Educational, Scientific, and Cultural Organization (UNESCO)

You got a problem, they got an agency monitoring and working in that area. They come out with status reports every now and then, detailing progress, and lingering problems in every area of human concern.

Global Think Tanks

Think Tanks play a powerful role in influencing policy in many critical areas through research and analysis. You can read their reports from their websites. Some of the most influential global think tanks [6.1, 6.2] are:

- Brookings Institution
- Chatham House
- French Institute of International Relations
- Centre for Strategic & International Studies (CSIS)
- Fraser Institute
- Amnesty International (AI)
- Carnegie Endowment for International Peace
- Transparency International (TI)
- Bruegel
- RAND Corporation
- China Institutes of Contemporary International Relations (CICIR)
- Wilson Center
- Fundacao Getulia Vargas (FGV)
- Council on Foreign Relations (CFR)

Country Ministries and Departments

Every country is governed through a number of Ministries or agencies with various departments under them. Each Ministry/ agency has specific departments that are tasked with the job of reviewing and framing the development policy related to the area assigned to it. These agencies release periodical reports assessing the impact of existing policies and suggesting a future course of action. Hence, they are an important source to understand the problems, which are currently plaguing a country. The reports of these agencies can be downloaded from their websites.

Browse these websites to understand the various agencies working under a government:

USA – www.usa.gov

UK – www.uk.gov

India – www.india.gov.in

Write down the critical problems, which you would like to solve after browsing through the above sources of information:

Local Problem (in your town):

Regional Problem (in your state):

National Problem:

Global Problem:

Inspiring Inventions

The boy who was considered unfit for school, but obtained over 1000 patents

Thomas Alva Edison (1847 – 1931) attended school only for a few months, as he was considered mentally ill and unfit to get an education by his teacher. The truth being that Edison was not interested in memorizing his lessons as was required by his teachers. Subsequently, he was educated at home by his mother. He read all the scientific magazines available at that time. As a teenage boy, Edison began selling newspapers, candy, and vegetables on trains and even employed other boys in his business. He used most of his earnings to buy scientific magazines, chemicals, and scientific apparatus. He set up a laboratory and conducted experiments in an unused luggage car in the train [6.3, 6.4].

STEP 7

Sustainability - the Holy Grail!

Earth provides enough to satisfy every man's need, but not every man's greed - Mahatma
Gandhi

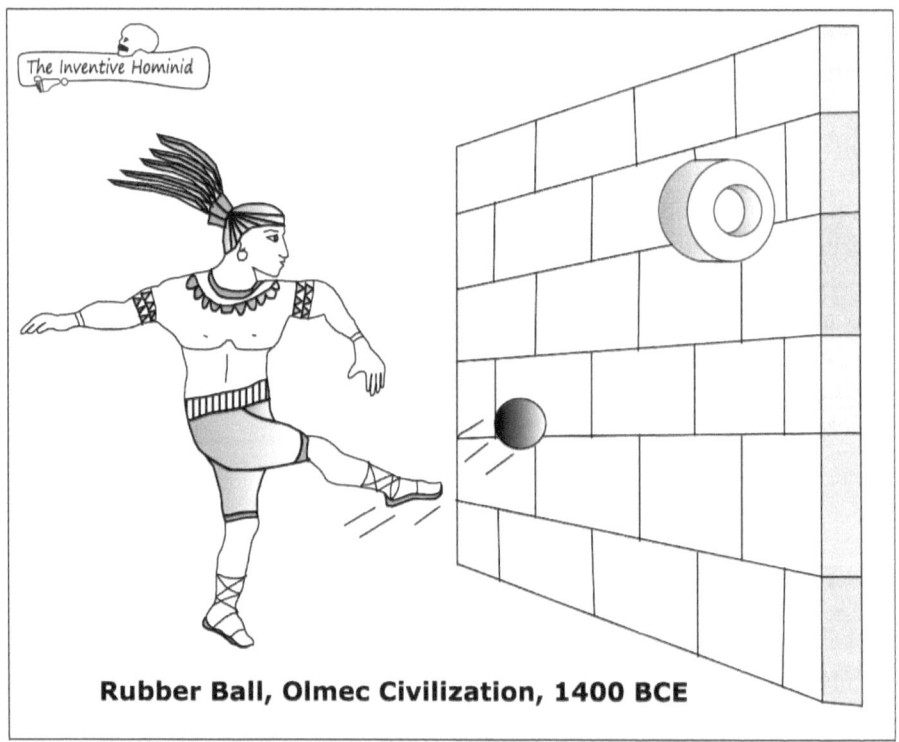

Rubber Ball, Olmec Civilization, 1400 BCE

TIH - Rubber Ball (7 IR 1, 7 IR 2, 7TR 1, 7TR 2)

Learning Objectives

After studying the contents of this chapter, you should be able to:
- Understand the concept of sustainability and sustainable development
- Know the sustainable development goals listed by United Nations
- List out sustainability goals which you want to follow and tackle in your life

Do you want to jump into the race for the noblest prize in problem solving? Try to provide a solution that is sustainable.

OK! OK! We get you; you want to know what is this beast called sustainable or sustainability? Here we go...

Sustainable Development

"Sustainable development is development that meets the needs of the present without compromising the ability of future generations to meet their own needs. It contains within it two key concepts: The concept of 'needs', in particular the essential needs of the world's poor, to which overriding priority should be given; and The idea of limitations imposed by the state of technology and social organization on the environment's ability to meet present and future needs."

-*Report of the World Commission on Environment and Development: Our Common Future, UN Bruntland Commission, 1987.*

Whoa! Easy, easy, are you seeing stars in your head. Good going, solve this and you are going to be a star in the future...

Solve a problem and provide a solution that is SUSTAINABLE!

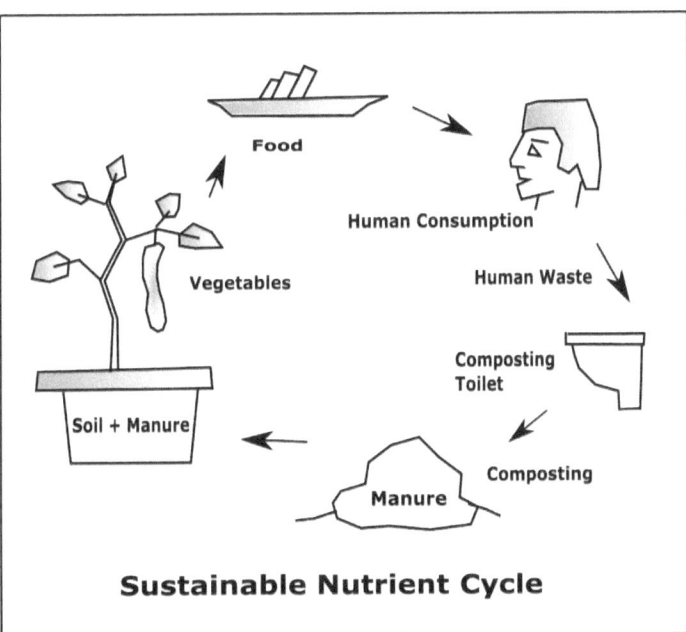

Sustainable Nutrient Cycle

Take the organic waste that is generated in your kitchen (vegetable peels, food waste, etc.), compost it, and grow vegetables with that. Sustainable to the core! Now apply this principle to the problem you have chosen to solve.

UN Sustainability goals

Want to be a savior of humankind by being a sustainability messiah? Here is the whole set of UN Sustainability goals [7.1]:

1 No Poverty	2 Zero Hunger	3 Good health & Well Being	4 Quality Education
5 Gender Equality	6 Clean Water & Sanitation	7 Affordable & Clean Energy	8 Decent Work & Economic Growth
9 Industry, Innovation & Infrastructure	10 Reduced Inequalities	11 Sustainable Cities & Communities	12 Responsible Consumption & Production
13 Climate Action	14 Life on Land 15 Life in Water	Peace, Justice & Strong Institutions	17 Partnerships for the Goals

To make a big dent on the efforts to save our earth, pick any one of the above-mentioned problems and jump right into it.

List down the sustainability goals, which you will make your life's mission and tackle it:

_____

_____

_____

_____

Earth will go on, even without humans, it is about saving ourselves

Human activities are linked to about 1.0°C of global warming above pre-industrial levels, with a likely range of 0.8°C to 1.2°C. Global warming is likely to reach 1.5°C between 2030 and 2052, if it continues to increase at the current rate. Warming above 1.5°C will result in major disruption in ecosystems around the world causing problems to our life supporting systems [7.2].

This can be avoided only by drastically altering our consumption of resources to sustainable rates and reducing our ecological footprint.

The Ecological Footprint is derived by tracking how much biologically productive area it takes to provide for all the competing demands of people. These demands include space for food growing, fiber production, timber regeneration, absorption of carbon dioxide emissions from fossil fuel burning, and accommodating built infrastructure. A country's consumption is calculated by adding imports to and subtracting exports from its national production.

The sustainable Ecological footprint has been calculated to be 2.1 global hectares/person (gha/person).The current ecological footprints of some countries are USA: 8.4 gha, China: 3.7 gha, India: 1.1 gha, Oman: 6.3 gha, Brazil: 3.1 gha [7.3].

Failure of humanity to reduce its resource consumption will lead to transformation of Mother Earth into a place inhospitable for humans. We can prevent it if every individual like you, becomes the change agent.

STEP 8

Looking For a Problem - Research Focus in Top Universities

A university is just a group of buildings gathered around a library. The library is the university - Shelby Foote

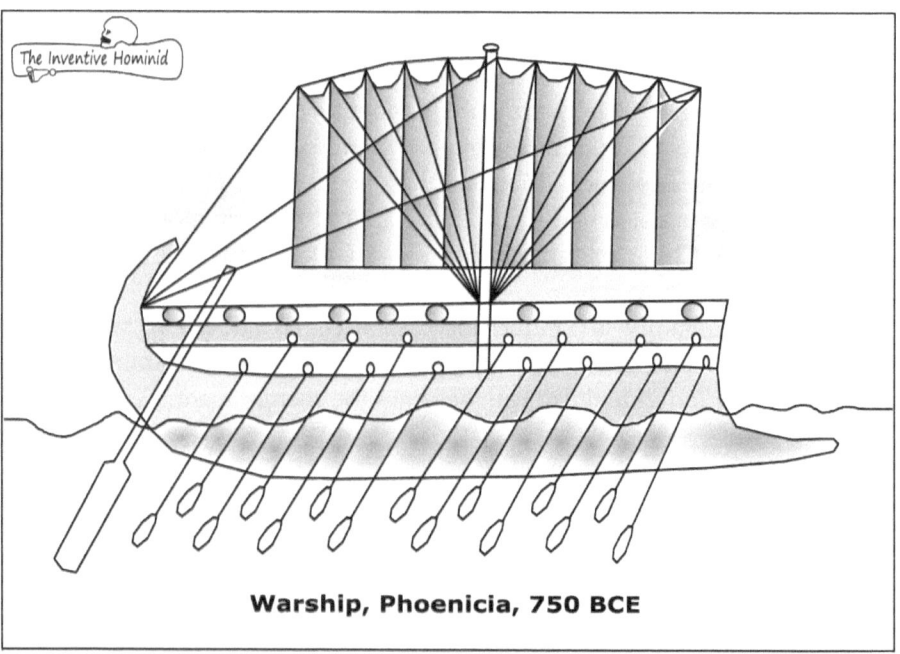

Warship, Phoenicia, 750 BCE

TIH - Phoenician Warship (8 IR 1)

Learning Objectives

After studying the contents of this chapter, you should be able to:
- Learn to access the current research happening at leading universities
- Explore and list down current research happening at leading institutes in the area of your interest.

You are very busy and want a quick way of finding out the critical problems instead of reading loads of UN reports.

Fine, we do have a cure for that.

Just log in to the research page of the top universities of the world and guess what, they have done all the spadework and have listed the most critical problems they are working on. Go ahead, have a free meal, help yourself.

MIT – Current research

Let us say we get into the website of MIT (web.mit.edu) and then navigate to the Research page, it lists all the institute research areas by topic.

A click on the area of Robotics and Artificial intelligence, lists about 10 labs / centers/ programs that come under it. Now when we enter into the page of Robot Locomotion Group, it gives information about their current research in terms of people, publications, lectures, competitions, etc.

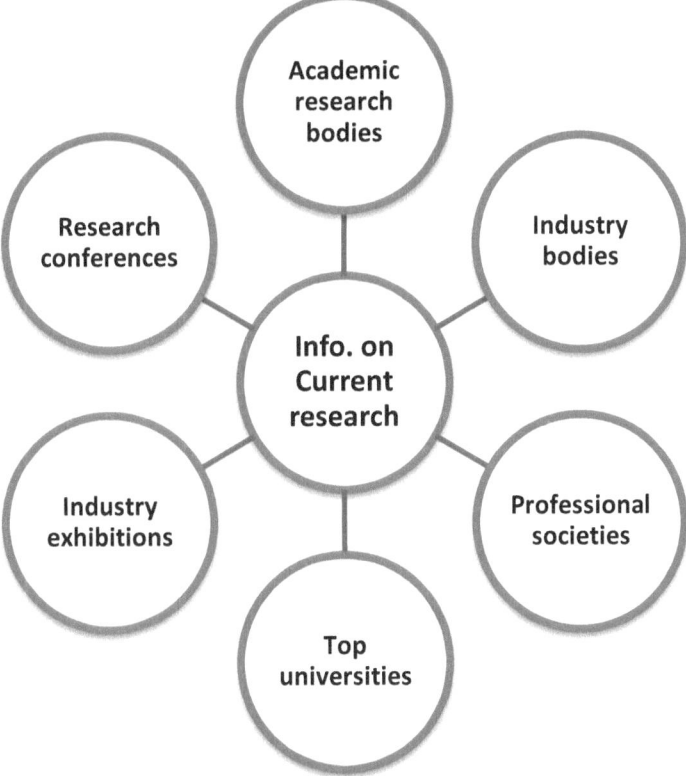

Cool, with a few mouse clicks, we have now understood the latest take on Robot Locomotion Research. Now just browse through the research areas of your interest and the labs within the whole of MIT and enjoy the sparks in your neurons as they sparkle with joy at the deluge of information.

Treat your brain in this similar manner by going through the research areas of top institutes in the world: Caltech, Stanford, Oxford, Cambridge, ETH Zurich, Princeton, NUS, and IITs etc. Just go through the whole list of top engineering schools in the world and fill your brain with the latest tech jargon.

List of research areas that you would like to explore more:
Institute:
Research Area:

Inspiring Inventions

The most important invention in the second millennium

There was a time when knowledge was available only to the elites of the society. It was so because the books at that time were handwritten and very expensive. A single copy of a book used to take a year for it to be completed. Hence, the invention that made books available to the masses is considered the defining invention of the second millennium.

The mechanical movable type printing press is that invention and the inventor was Johannes Gutenberg (1400-1468). Gutenberg was born in the German city of Mainz in a family of merchants and Goldsmiths. Gutenberg in 1439 used the movable type for the first time. His contributions that revolutionized printing are a process for mass-producing movable type; use of oil-based ink; adjustable molds and mechanical movable type [8.1, 8.2].

This innovation introduced the mass production of books and the age of mass communication. It encouraged unrestricted circulation of revolutionary ideas, increased literacy, made knowledge available to everybody, and ultimately ushered in democracy [8.3].

Gutenberg did not benefit financially in his lifetime, but the printing technology spread across Europe, scientific ideas in the form of books reached all corners of Europe and was the main catalyst for the scientific revolution.

STEP 9

Looking For a Problem - Industry Bodies

'God forbid that India should ever take to industrialism after the manner of the West. ... If an entire nation of 300 million took to similar economic exploitation, it would strip the world bare like locusts'. -Mahatma Gandhi

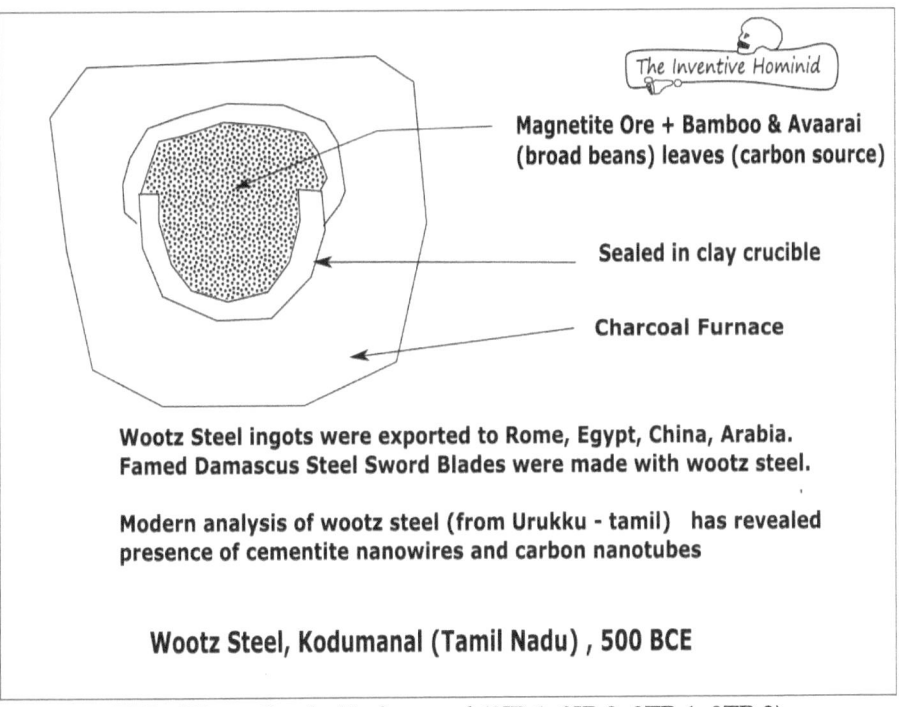

The Inventive Hominid

Magnetite Ore + Bamboo & Avaarai (broad beans) leaves (carbon source)

Sealed in clay crucible

Charcoal Furnace

Wootz Steel ingots were exported to Rome, Egypt, China, Arabia. Famed Damascus Steel Sword Blades were made with wootz steel.

Modern analysis of wootz steel (from Urukku - tamil)   has revealed presence of cementite nanowires and carbon nanotubes

Wootz Steel, Kodumanal (Tamil Nadu) , 500 BCE

TIH - Wootz Steel - Kodumanal (9IR 1, 9IR 2, 9TR 1, 9TR 2)

Learning Objectives

After studying the contents of this chapter, you should be able to:
- Understand the role of industry bodies in promoting research and best practices
- Identify the industry body in the area of your interest
- Explore the reports released by them and understand the current advances

Industry bodies (Associations/ Organizations)

Industry bodies provide information on current problems in their domain. Every field will have an industry body that sets the tone in terms of best practice benchmarks, code regulations, etc. some of the well-known industry bodies are:
- World Steel Association
- World Foundry Organization
- International Fertilizer Industry Association
- World Gold Council
- World Nuclear Association
- International Federation of Robotics

The websites of these bodies provide the entire information spectrum associated with their industry in terms of statistics, technical reports, manuals, conferences, and competitions.

Become a student member of these associations and get a head start in becoming a successful industrial professional by researching into the current problems the industry in facing.

Note down the industry bodies connected with your area:

Current problems they are researching:

Inspiring Inventors

The woman who stopped bullets, that is still saving many lives

Stephanie Luoise Kwolek (1923-2014) was an American chemist best known for inventing poly-paraphenylene terephthalamide – popularly known as Kevlar [9.1]. Born to polish immigrant parents, her father died when she was ten years old. She earned her bachelor degree in chemistry and hoped to work until she had saved enough money to study medical course to become a doctor.

She was offered a job at DuPont, which was vacant due to men being away overseas in World War II. Finding the work interesting, she stayed in that job and after ten years, she discovered Kevlar. It was an accidental discovery. While she was searching for a lightweight yet strong fiber for use in tires, one of her combinations produced buttermilk like solution. Instead of throwing it away, she convinced the technician to run the spinneret test on her solution. She was amazed to find that the new fiber was stronger than nylon and five times stronger than steel by weight [9.2].

Thus, she discovered Kevlar, which is used to make bulletproof vests apart from applications such as parachute lines, aircraft parts, car tires, fire fighting boots, bombproof materials, armored cars, etc. Her discovery generated billions of dollars in revenue for DuPont and she was inducted into the National Women's Hall of Fame in 2003[9.2].

STEP 10

Looking For a Problem - Professional Associations

Everyone has the right to freedom of peaceful assembly and association.
-Article 20 of the Universal Declaration of Human Rights

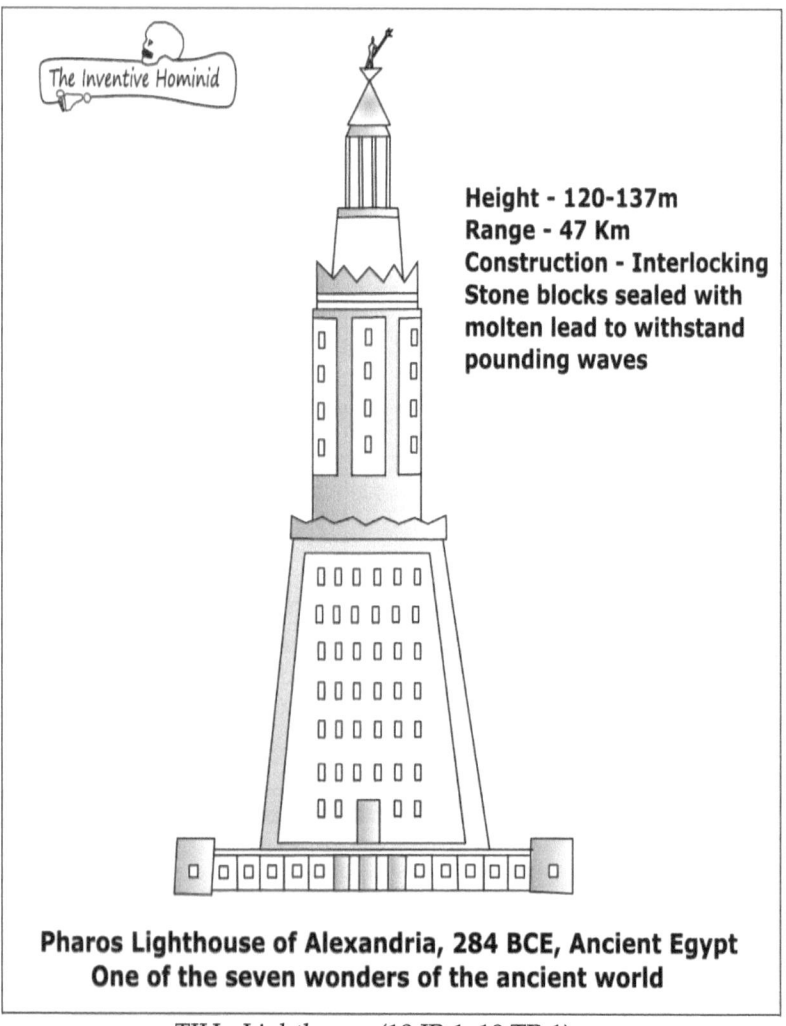

The Inventive Hominid

Height - 120-137m
Range - 47 Km
Construction - Interlocking
Stone blocks sealed with
molten lead to withstand
pounding waves

Pharos Lighthouse of Alexandria, 284 BCE, Ancient Egypt
One of the seven wonders of the ancient world

TIH - Lighthouse (10 IR 1, 10 TR 1)

Learning Objectives

- After studying the contents of this chapter, you should be able to:
- Understand the role of professional associations
- Identify professional associations of your interest
- Find membership information and participate in conferences

Another avenue to keep abreast of what is happening in your engineering area is by logging onto the websites of professional organizations. Every branch and sub-branch of engineering has professional organizations.

Professional Organizations

- Association of Enterprise Architects (AEA)
- Association of Information Technology Professionals (AITP)
- The Institution of Engineers (India)
- American Society of Civil Engineers (ASCE)
- Association for Computing Machinery
- Institute of Electrical and Electronics Engineers (IEEE)
- Indian Institute of Chemical Engineers
- Institute of Food Technologists
- Aeronautical Society of India
- Computer Society of India
- Institute of Marine Engineering, Science, and Technology (IMarEST)
- Institute of Mechanical Engineers (IMechE)
- Institution of Chemical Engineers (IChemE)
- International Association for Bridge and Structural Engineering (IABSE)
- International Council on Systems Engineering (INCOSE)
- International Society for Pharmaceutical Engineering
- Project Management Institute (PMI)
- Society for Human Resource Management (SHRM)
- Society of Automotive Engineers (SAE International)
- Society of Manufacturing Engineers (SME)
- Society of Tribologists and Lubrication Engineers (STLE)
- World Federation of Engineering Organizations (WFEO)

These organizations conduct yearly conferences, which provide registration for students at a discounted rate. Try to attend these conferences and network among the professionals in your area, who knows you might land a prized internship and/or a job offer.

List down the names of professional associations in the area you are interested in:

_____

Inspiring Inventions

Do you enjoy canned food? Well, thank the Father of Canning

Nicolas Appert (1749-1841) was the French inventor of airtight food preservation. Appert was a confectioner and chef in Paris and began experimenting with ways to preserve various foodstuffs. His method consisted of placing food stuff in glass jars, sealing it with cork and sealing wax and placing it in boiling water, until the contents are thoroughly cooked [10.1, 10.2]]. In 1800, Napoleon offered a prize of 12,000 francs for a new method of preserving food mainly to be used by the French army. Appert presented various fruits and vegetables preserved in bottles at the Industrial Expo in 1806. He was given the reward of 12,000 francs in 1810 [10.1], after he published his method in a book. His invention spread rapidly and evolved into preservation in tin cans giving rise to the name canning. According to USDA, low-acid canned food such as meat, fish and vegetables can last up to 5 years.

Thus, this pioneering effort in food technology gave rise to a big industry that reduced wastage of food and made it available to the masses around the world.

STEP 11

LOOKING FOR A PROBLEM – ACADEMIC/ SCIENTIFIC RESEARCH BODIES

One should observe our scholars closely: they have reached the point where they think only "reactively," i.e. they must read before they can think.
-Friedrich Nietzsche, The Will to Power

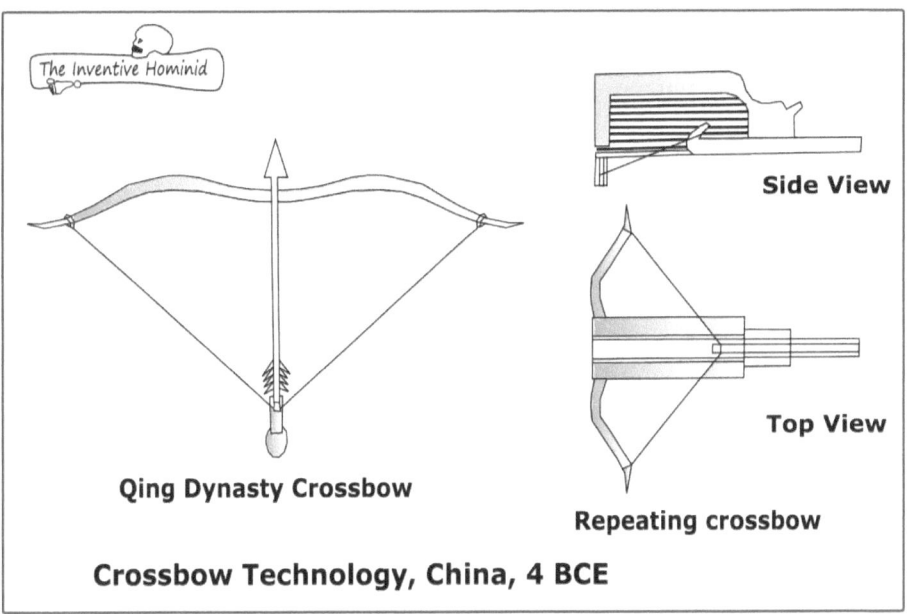

The Inventive Hominid

Qing Dynasty Crossbow

Side View

Top View

Repeating crossbow

Crossbow Technology, China, 4 BCE

TIH - Crossbow Technology (11 IR 1, 11 TR 1, 11 TR 2, 11 TR 3)

Learning Objectives

After studying the contents of this chapter, you should be able to:
- Understand the role of academic research bodies
- Identify research bodies in the area of your interest
- Explore the current advances in the societies/ bodies of your interest

If you have an academic bent in your aptitude and want to get into the academic profession, then better get acquainted with the prestigious academic bodies in your field.

When we talk about academic research and teaching, the distinction between science and engineering gets narrow. Fundamental contributions in any field have to be rooted in the basic and mathematical sciences. These bodies are mostly centuries old tracing their history back to the great scientists of yore.

Whatever might be your branch of engineering, the basic concepts are always rooted in the physical, chemical, and mathematical sciences. This also gives you an opportunity to pursue your passion in pure sciences that you might have experienced in your school days. Some of the academic bodies that you can explore are:

Academic/ Scientific Research Bodies

- Association for Symbolic Logic
- Human Behavior and Evolution Society
- International Academy of Quantum Molecular Science
- Chemical Research Society of India
- International Association for Mathematics and Computers in Simulation
- International Behavioral Neuroscience Society
- International Behavioral and Neural Genetics Society
- Crpytology Research Society of India
- International Communication Association
- International Council for Philosophy and Humanistic Studies
- International Council for Science
- Indian Mathematical Society
- International Institute for Conservation
- International Mammalian Genome Society

- International Mathematical Union
- International Social Science Council
- International Union of Microbiological Societies
- Plasma Science Society of India
- International Union of Pure and Applied Chemistry
- Society for Animation Studies
- Society for Social Studies of Science

List down the names of academic research bodies in the area you are interested in:

Inspiring Individuals

Nobel Laureate Venkatraman Ramakrishnan

Venkatraman Ramakrishnan was born in Chidambaram (India) in 1952. He is an American and British Structural biologist of Indian Origin. He obtained his education from MS University, University of California, and Ohio University (PhD). After his post-doctoral fellowship, he could not get a faculty position even after applying to about 50 Universities. In 1999, he started his current position in Cambridge, England. In 2009, he won the Nobel Prize in Chemistry along with Thomas Steitz and Ada Yonath for studies of the structure and function of the ribosome [11.1, 11.2].

Venkatraman Ramakrishnan is currently the president of **The Royal Society of London**, the oldest scientific institution in the world founded in 1660. Scientific stalwarts like Isaac Newton, JJ Thompson, and Stephen Hawking have been fellows of this society [11.2].

STEP 12

LOOKING FOR A PROBLEM - THEMATIC AREAS OF UPCOMING CONFERENCES

When Demosthenes was asked, what was the first part of Oratory, he answered, "Action," and which was the second, he replied, "Action," and which was the third, he still answered "Action." -Plutarch, Morals. Lives of the Ten Orators.

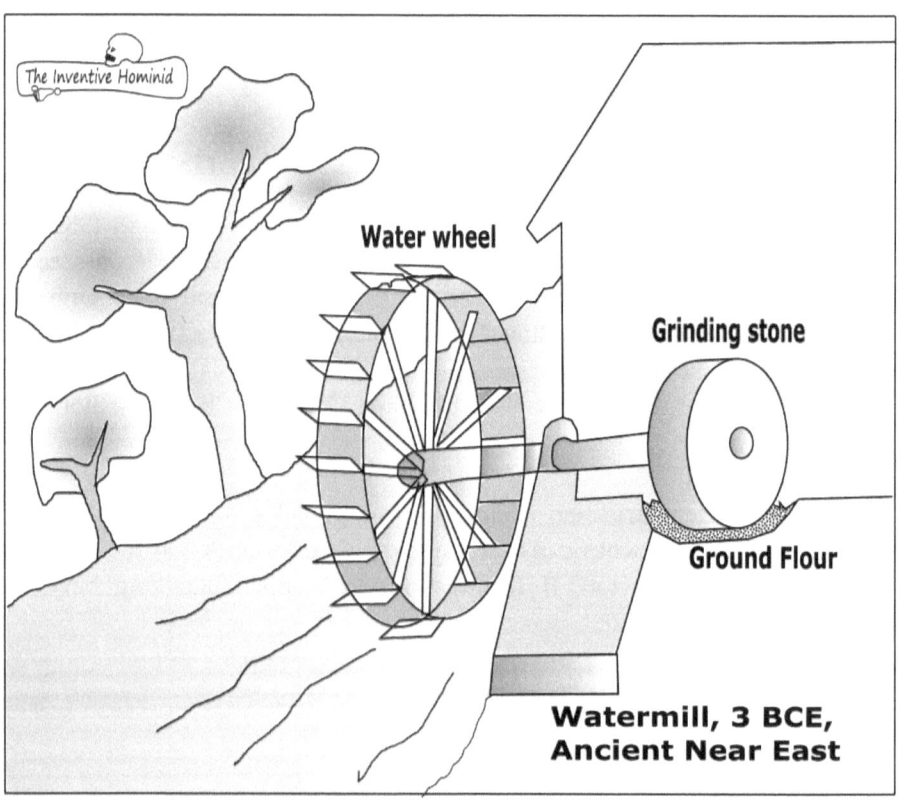

The Inventive Hominid

Water wheel

Grinding stone

Ground Flour

Watermill, 3 BCE, Ancient Near East

TIH - Water Wheel - Water Mill (12 IR 1, 12 TR 1)

Learning Objectives

After studying the contents of this chapter, you should be able to:

- Understand the role of conferences in advancing knowledge and collaboration
- Identify upcoming conferences in your area and register in them
- Access the online proceedings and abstracts of past conferences and explore them
- Setup an email alert for your favorite conferences

To get an idea of what is currently buzzing in the engineering circles, a glance at the upcoming conference themes of top industry, academic and professional bodies will give a good idea.

CONFERENCE ALERTS

Set up a conference alert on your favorite area of engineering and be updated with the regular stream of alert emails. Most conference websites will provide the list of abstracts of papers received and full papers after the conference are over. These abstracts and papers give an inside glimpse on what is cooking in the top labs around the world.

PROCEEDINGS OF PAST CONFERENCES

Proceedings are compilations of abstracts, papers, and presentations given at a particular conference. Just browse through the proceeding of past conferences and just soak up the entire gamut of information to enrich your knowledge in the area of your interest.

List down the conferences that you would like to track for the next one year:

Title of Conference	Organized by	Dates

The Great Exhibition of 1851

Great Exhibition of the works of industry of all nations was held in London. A special building made of cast iron frame and glass measuring 563 meters long by 138 meters wide housed the exhibition featuring cultural products and Industrial inventions. The building enclosed fully-grown trees and was an engineering and architectural marvel by itself. The exhibition ran for 168 days and 6,039,722 people visited. The exhibition generated a profit of 18 million pounds (2015 value) which was used to build three museums and establish a trust to provide scholarships for industrial research, which continues until today. Products were exhibited from Britain's colonies and 44 foreign countries. Some of the inventions and artifacts exhibited were the Jacquard Loom, an envelope machine, kitchen appliances, steel making displays, reaping machine, Kohinoor and Darianoor diamonds, fax machine, photography, revolvers, etc. The event featured first modern pay toilets which charged 1 penny to use them [12.1, 12.2].

STEP 13

STAYING UPDATED ABOUT PROBLEMS

"We don't know a millionth of one percent about anything."-Thomas Edison

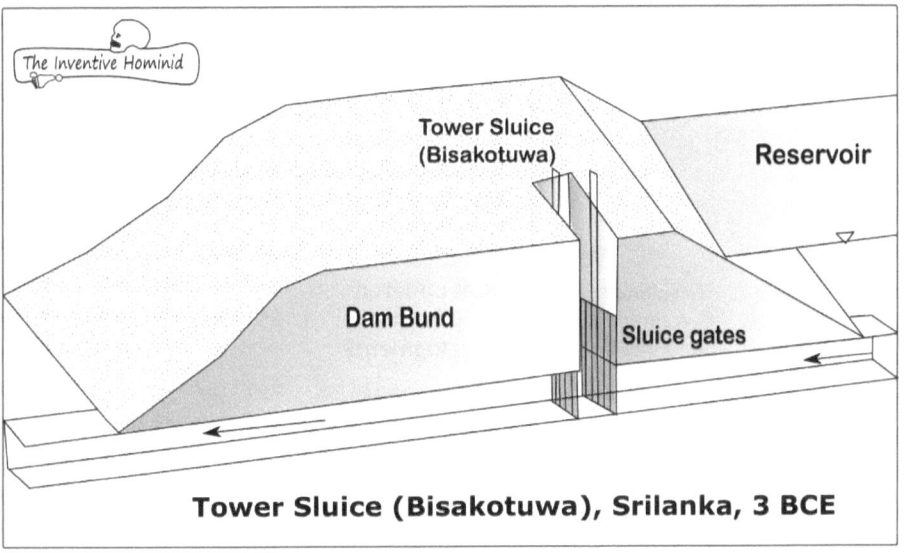

Tower Sluice (Bisakotuwa), Srilanka, 3 BCE

TIH - Bisakotuwa - Tower Sluice (13 IR 1, 13 TR 1)

Learning Objectives

After studying the contents of this chapter, you should be able to:
- Understand ways to stay updated about developments
- Subscribe to e-newsletters
- Set up alerts in news and social media about problems of your interest

How frequently should you look up the websites of professional, academic and industry research bodies to stay updated?

E-NEWSLETTERS

An easy way out is to subscribe to newsletters of these bodies, these newsletters will contain information about latest developments and will be delivered to your email as and when they are published, daily weekly or monthly.

Want a faster way to stay abreast with the updates in your favorite websites?

SUBSCRIBE TO RSS FEEDS.

Most websites allow users to subscribe to their RSS feed. RSS feeds will give the subscriber a list, which will reflect the latest items that have been updated on a website. Therefore, you can save time by reading the latest updated items, instead of wasting time by scanning the entire website for changes.

WEB ALERTS

Well, we all want to be the first to know about developments in anything about our interest. A cool way to have information flowing to you as it happens on the internet is to subscribe to web alerts. A good place to start can be the Google alerts, which is free. Use your Google account to set up any number of alerts. Just set up an alert for a key word of your interest say: fuzzy logic and adjust the setting to deliver the result to your inbox. You can set it to alert you as it happens, daily or weekly. The results will be delivered to your inbox as per the frequency you have selected. There are a number of alternatives to Google alerts, which you can use if you feel that you want more than what Google can give you.

TECH UPDATES IN THE SOCIAL MEDIA

Why vilify social media for all the negative aspects? There are positive aspects that we can utilize. In Facebook, you can like pages related to the technology of your interest and when you access Facebook, directly go to pages feed to soak up the latest technology news.

Inspiring Inventions

The net in which we are all entangled

The Internet is the worldwide system of interconnected computer networks using the Internet protocol suite (TCIP/IP) to link a wide range of devices. The linkage is provided by an array of electronic, wireless, and optical networking technologies. Its origin date back to the research sponsored by US Government in the 1960s to build fault-tolerant communication with computer networks [13.1]. Common methods of internet access include dial-up modems, broadband, fiber optics, Wi-Fi, satellite and cellular telephony (4G, 5G). It is estimated that around 4 billion people will be using the internet by 2020. The advent of the internet has revolutionized online education, social networking, healthcare, entertainment, banking, governance, disaster management, etc.

The reach of the internet in our daily lives will explode in the coming years through the convergence of technologies such as AI, machine learning, image processing, biometric surveillance, IOT, etc.

STEP 14

APPROACHES/WAYS OF SOLVING A PROBLEM

"A problem never exists in isolation; it is surrounded by other problems in space and time. The more of the context of a problem that a scientist can comprehend; the greater are his chances of finding a truly adequate solution".
- Russell L. Ackoff (1956)

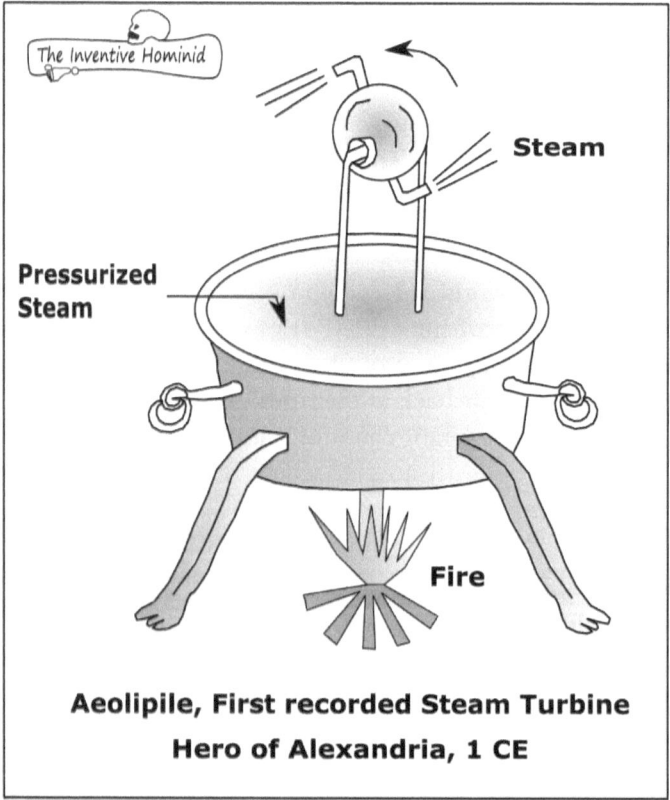

Aeolipile, First recorded Steam Turbine
Hero of Alexandria, 1 CE

TIH - Steam Turbine (14 IR 1, 14 TR 1)

Learning Objectives

After studying the contents of this chapter, you should be able to:
- Learn about various ways to contribute to development of knowledge
- List out which approach you would like to take in your project

Now that you are ready to help humanity by tackling a critical problem, what do we mean by solving a problem? Does only inventing a new product count as a contribution? What are the ways one can contribute to solve a problem?

A project aimed at solving a problem can be of many types.

APPROACHES TO SOLVING PROBLEMS

PRODUCT DEVELOPMENT

Are you the type who spent hours and hours in your childhood taking apart toys to understand how they worked? Then developing a product by combining various elements to create a working miracle might be the path for you.

It can involve developing a new product to solve a problem or satisfy a demand. Developing a fitness band is an example of fulfilling a need in the fitness arena. Development of small family scale Reverse osmosis based water filter is an example of a product that has served a critical need for drinking water in areas with saline ground water.

DESIGN IMPROVEMENT

A project can also improve upon existing designs by incorporating advances in technology into existing products. A smart phone is an example of a design improvement in feature phones and what a revolution it has been.

PROCESS IMPROVEMENT

Process improvement pertains to the improvement in complicated processes involved in accomplishing certain results such as manufacturing a car or a candy or an aircraft or a cookie. Small improvements in these complex processes such as introduction of automation or sensors can result in tremendous increases in productivity.

THEORY DEVELOPMENT

Theory refers to the fundamental science behind certain concepts. For example, you can delve into the theory of heat transfer across various materials or derive a more efficient expression to capture the lift and drag forces acting on bodies moving across a fluid. If you are the type who relishes in wrestling with problems involving complex mathematics or conceptual conundrums, then theory development is the way to go.

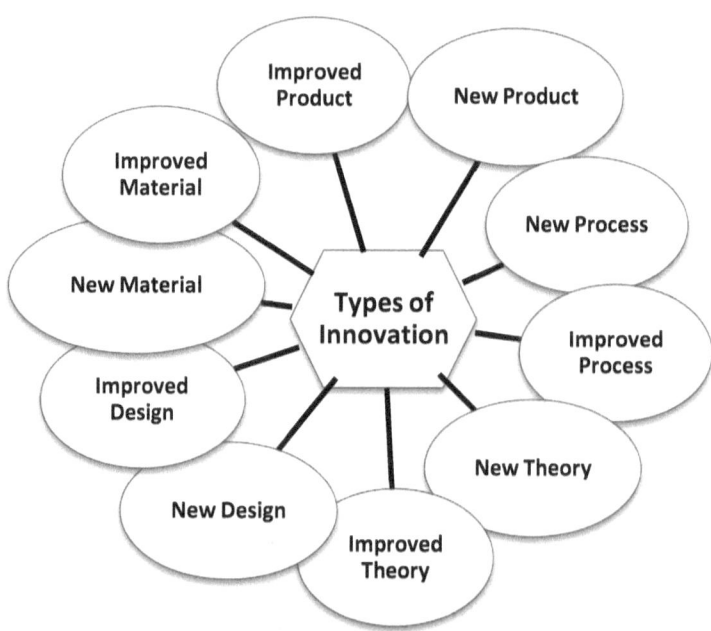

NEW MATERIALS, COMPOSITES, ALLOYS, BIOMIMICRY

Would you like to join the hottest pursuit in our planet? It is the race to find the super materials that can super conduct heat, withstand high temperatures, and do much more. Imagine any product, structure or a machine. It is essentially made of materials that have been chosen for their unique properties. Now when we come up with better materials, it results in a tumultuous change in the performance of that product such as more efficiency, lesser weight, more functionality, etc. A good example is the evolution of mobile phones from being a brick size object to a tiny matchbox type model. Alternatively, you can study beings in nature and mimic their amazing capabilities to create a new technology, which is called biomimicry or biomimetics [14.1].

Write down the approach that you would like to take in your project:

Inspiring Inventors

What would you do if your business were destroyed twice?

Soichiro Honda (1906 – 1991) was the Japanese founder of Honda Motor Co. Ltd. Honda in his childhood helped his Father in his Bicycle repair business. He left his home at the age of 15 and worked as a car mechanic for six years in Tokyo. He returned home at the age of 22 and started an auto repair business. In 1937, Honda started two factories to produce piston rings for Toyota. However, fate had other plans; the Yamashita factory was destroyed by a bomb from an US B-29 and the Iwate factory collapsed in an earthquake. He sold the salvageable items to Toyota and with the money started the Honda Research Institute. In 1948, he started producing a complete motorized bicycle and in 1949 produced a complete 2-stroke motorcycle. Thus began the journey in which Honda turned it into a billion dollar company, which dominated even Triumph and Harley-Davidson in their home countries [14.2, 14.3].

STEP 15

WHICH DIRECTION IS THE WORLD OF INNOVATION HEADED?

"The key to success for Sony, and to everything in business, science, and technology for that matter, is never to follow the others."
-Masaru Ibuka, founder of Sony

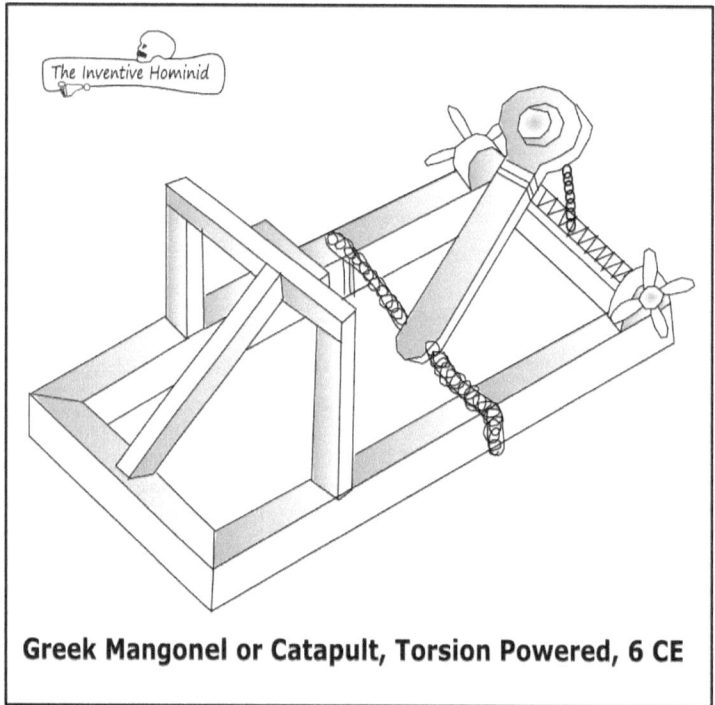

Greek Mangonel or Catapult, Torsion Powered, 6 CE

TIH - Catapult (15 IR 1, 15 TR 1)

Learning Objectives

After studying the contents of this chapter, you should be able to:

- List the areas the world of innovation is currently focusing on
- Identify areas which is of your interest, for deeper understanding

Currently we are in the midst of a revolution involving AI. We are trying to build things, which can think and make decisions like us. In short, we are trying to replace as many human workers as possible with machines that can do the job without getting tired or without asking for a salary raise or vacation.

Will this lead to a situation where machines will do everything and leave us with unlimited leisure? As envisaged by Herbert Simon?

We are talking about a world full of intelligent things ranging from our toasters, cars, bricks, walls, lights, ovens, etc., everything communicating with each other and dancing to our preferences.

THE TECHNOLOGIES AT THE FOREFRONT OF FUTURE RESEARCH

The technologies that are at the forefront of future research are [15.1, 15.2, 15.3]:

- Artificial Intelligence
- Driverless Vehicles
- Automated Surgery
- Humanoid Robot
- Artificial Photosynthesis
- Natural Language Processing
- Machine Learning
- Arterial Nanobots
- Gene Editing/ Therapy/ Stitching/ Design
- Interplanetary/ Interstellar Mobility/ Habitat Design
- Sustainable Mobility
- Brain Transplant
- Designer Antibiotics
- Carbon Sequestration
- Vegan Meat
- Sustainable Concrete/ Smart Buildings
- Bacterial Fuel/ Solar Cells
- Wearable Electronics/ Diagnostic/ Smart Devices
- Microbial Computing/ Quantum Computing
- Genetically Modified Food/ Life Forms
- Sustainable Off Grid Energy/ Wireless Charging
- Virtual/ Augmented Reality
- Advanced Vaccines/ Drug Delivery/ Anti-Ageing Therapy

- Bit-Coin/ Block Chain Technology
- Artificial/ 3D Printed Organs
- Animal Grown Critical Organs for Human Transplant
- Mining In the Moon/ Planets/ Asteroids
- Nanotechnology - Nano Sensors
- Advanced 2D Materials Similar To Graphene
- Drones / Unmanned UAVs

Future Technologies

List areas that arise in your mind for deeper exploration:

Inspiring Inventors

The invention that made books available to the masses.

Friedrich Gottlob Keller (1816 – 1895) was a German Machinist who invented the woodcut machine used to extract fibers needed for wood pulp process of papermaking. In those times, paper was made with waste cloth. Unsatisfied with working as a weaver along with his father, he began devouring science publications dealing with machinery. After 3 years of hard work, he finally perfected the woodcut machine and produced paper from wood pulp. Unable to get funding from the Government to develop paper making further, he sold his invention to paper manufacturer Heinrich Voelter for 80 Pounds. A patent was granted in 1845 in the names of Keller and Voelter. In 1952, when the patent came up for renewal, Keller did not have the money to renew his part. Hence, Voelter become the sole patent holder without Keller and earned huge profits [15.4, 15.5].

As his invention became popular and revolutionized papermaking, Keller became unemployed and penniless. However, in 1870 German paper makers and their associations collected money and gave him enough to buy a house and worry free retirement.

STEP 16

ARTIFICIAL INTELLIGENCE

A computer would deserve to be called intelligent, if it could deceive a human into believing that it was human- Alan Turing, Computing Machinery and Intelligence (1950)

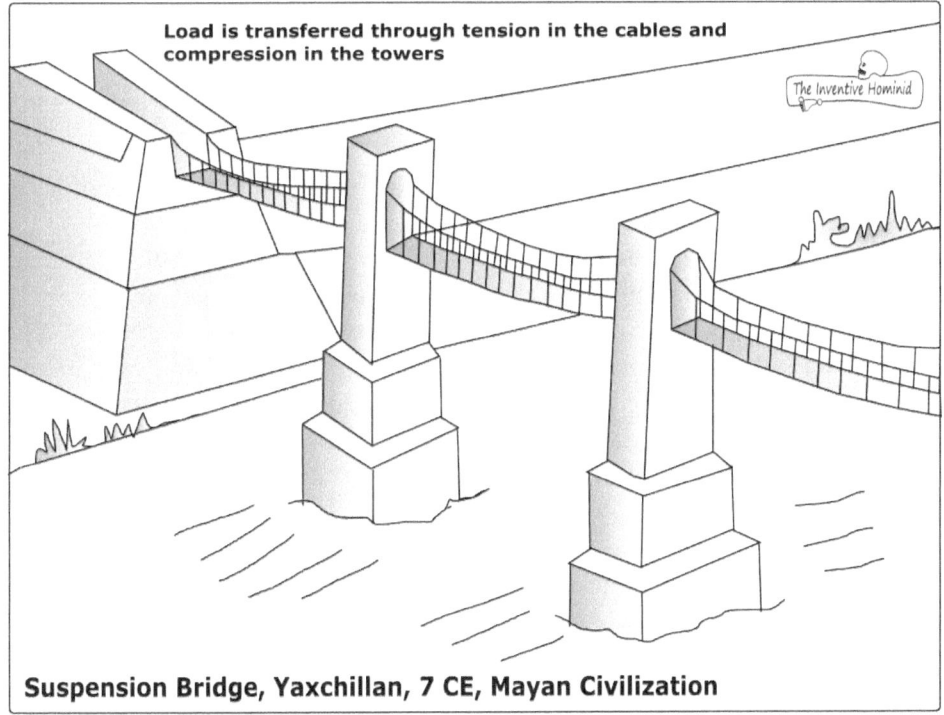

Load is transferred through tension in the cables and compression in the towers

The Inventive Hominid

Suspension Bridge, Yaxchillan, 7 CE, Mayan Civilization

TIH - Suspension Bridge (16 IR 1, 16 TR 1)

Learning Objectives

After studying the contents of this chapter, you should be able to:
- Know about the exciting area of AI and its allied areas
- Know about areas that employ AI for its advancement
- Know about online resources available for learning more about AI

Artificial intelligence as the name suggests, involves capturing the process of human thinking and decision making to enable a machine to do the same. The human brain comprises of 100 billion neurons, each firing 200 times per second. Therefore, we can understand the enormity of the task involved.

Regardless of your specialization, if there is one area that can give your career a massive boost and enable you to make a mark on our world, it is AI. In the future, companies and individuals without any expertise in AI will be relegated to the bottom.

Due to enormous advances in computing power, it is going to be a struggle for humans to hold on to their jobs against the onslaught of AI powered smart machines.

AREAS WHERE AI IS PLAYING A ROLE 16.1, 16.2, 16.3

Machine Learning - to improve algorithms automatically through iterations of learning.

Natural Language Processing - gives the machines, the ability to take commands directly in natural languages of humans as opposed to specialized programming languages that requires a high level of expertise.

Robotics - ever wondered how we humans are able to stand and walk on two legs in a balanced manner. We do it in an instinctive manner that involves complex coordinative messages between various muscles, nervous system, and our brain. Now replicating this complex aspect in robots with AI is an exciting field of research.

AI General Intelligence - refers to the area that is trying to create machines with reasoning and decision making capacity equal to or exceeding humans. This area also encompasses aspects such as Artificial Consciousness or Artificial Brain, an exciting realm for young researchers to make their mark.

Machine Perception - which is about machines using various inputs from sensors, cameras and other devices and be able to perform tasks like speech and facial recognition.

Knowledge Engineering - involves storing and retrieving information about a singular or a variety of subjects in vast databases and then teaching the machine to process it and come up with useful knowledge. Something similar to trying to duplicate a human who makes decisions about a particular situation based on years of information he has stored in his brain that we call as experience or wisdom.

Tools/ Approaches that are used in the field of AI, Which you can learn, include:

- o Cybernetics
- o Tools for AI
- o Cognitive Mapping/ Systems
- o Neural Networks
- o Fuzzy Logic
- o Bayesian Networks

CURRENT APPLICATIONS OF AI

(brief list): automated stock trading, facial/speech/pattern recognition, automatic gear boxes, air combat training, data mining, automated image processing, automated disease diagnosis, news composition, music composition, story writing, Driverless Cars of Google/ Tesla, Virtual Assistants Such As Alexa, etc.

EXCITING INFO-TECH THAT YOU CAN JUMP INTO FOR AN EXHILARATING CAREER

- o Advanced Machine Learning
- o Blockchain - Distributed Electronic Ledger
- o Home/ Office Automation
- o Brain - Computer Interface/ Man-Machine Interface
- o Digital Platforms/ Digital Currency
- o IOT/ IOE Platforms, Gesture Recognition, Exascale Computing, DNA Computing, Li-Fi, Nanoradio, Optical Computing, Subvocal Recognition
- o Quantum Computing - Using Quantum Mechanics
- o AI & IOT Convergence
- o Real Time Application Management
- o Embedded Cyber security
- o Dynamic Digital Tech
- o Geofencing
- o Advanced Cloud and Edge Computing
- o Big Data & AI Convergence
- o 3D Printing
- o Advanced Augmented and Virtual Reality
- o Deep Learning for Visual Processing
- o Open AI System - Collaborative AI Platform to Share Knowledge and Advances
- o Information Privacy
- o Social Analytics
- o Digital Business Models and Experimentation
- o Coursera-Machine Learning
- o Udacity-Intro to Machine Learning
- o Udacity-Deep Learning
- o Google -Google AI
- o OpenAI -openai.com
- o Deepmind

o Baidu Research
o AI2 - Allen Institute for Artificial Intelligence

 **LEARNING RESOURCES FOR AI**

Inspiring Inventors

The Man who cracked Enigma and saved 14 million lives

Alan Mathison Turing (1912-1954) is widely regarded as the father of theoretical computer science and Artificial Intelligence. His father, Julius Mathison Turing worked on the Indian Civil Service (ICS) at Chatrapur, Orissa. Turing was born and educated in England at the age of 13, unable to find transportation due to a general strike and determined to attend the first day of school, he rode his bicycle alone for 97kms to the boarding school. Turing was naturally inclined to mathematics and science much to the dismay of teachers concerned at providing a well-rounded education [16.4, 16.5].

Turing went on to study mathematics at King's college, Cambridge, and published path breaking papers on algorithms on computing machines. During the Second World War, Turing developed an electromechanical machine called the Bombe. This machine helped break the encrypted messages from the German Enigma machine, which was considered undecipherable at that time. This contribution enabled the allies to intercept coded German messages and defeat them in crucial battles. This ability by the allies shortened the war in Europe by more than two years and saved over 14,000,000 lives.

STEP 17

HOUSING

"If no serious action is taken, the number of slum dwellers worldwide is projected to rise over the next 30 years to about 2 billion." -Kofi Annan, Everyone Needs a Home

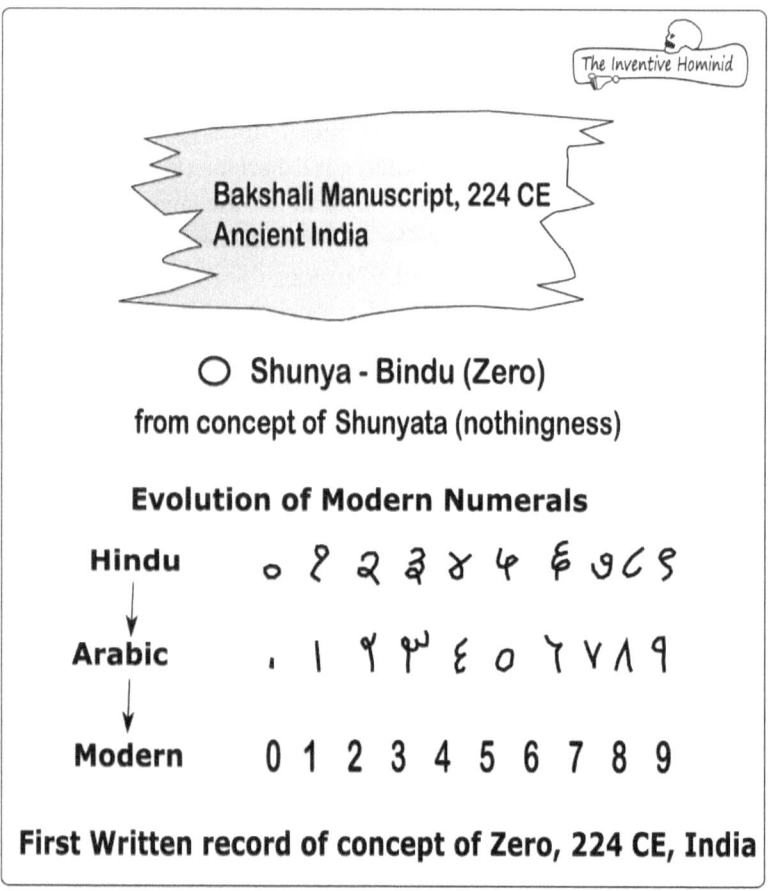

The Inventive Hominid

Bakshali Manuscript, 224 CE
Ancient India

○ Shunya - Bindu (Zero)
from concept of Shunyata (nothingness)

Evolution of Modern Numerals

Hindu

Arabic

Modern   0 1 2 3 4 5 6 7 8 9

First Written record of concept of Zero, 224 CE, India

TIH - Zero (17 IR 1, 17 TR 1)

Learning Objectives

After studying the contents of this chapter, you should be able to:
- Understand the categories of housing
- Current trends in housing and construction technology
- Learn about the significant link between climate change and housing

Housing is a necessity after food and clothing. Humans are spending a major portion of their life's earnings to build a house. The type of housing varies widely dictated by the climate and economic development of a region.

Housing can be categorized into these types:
- Built using only naturally available local materials
- Built with a combination of local and industrially produced materials
- Build predominantly with industrially produced materials

Hence, in the northern hemisphere rich with timber, houses are built with timber as a major component. In the tropics, it is changing from mud walls with vegetation roofing to brick walls with concrete or metal roofing. The economically impoverished areas are still struggling to provide proper housing.

HOUSING AND CLIMATE CHANGE

Housing contributes a huge amount of carbon dioxide emissions through its massive use of concrete and steel. Any effort to reduce the carbon dioxide emissions to halt climate change has to find ways to reduce the use of concrete and steel in buildings. Efforts directed towards a lower carbon footprint building will have a big impact in our efforts to halt climate change.

THE LATEST TRENDS IN HOUSING

Sustainable housing, which includes energy and resource efficient housing concerned about the embodied energy of building materials used in construction and the energy and resource usage in its day-to-day running.

Passive House is a building, which conforms to ultra-low German energy efficiency standard of Passivhaus. It typically requires that the total primary energy consumption must not be more than 60 kWh/sq.m/year.

Affordable housing concerned with providing housing for the multitude of ever-increasing people in the low-income category.

Tiny housing which aims to cut down on the size of the house and possessions to reduce debt and footprints.

Skyscrapers where land availability is very low and the land prices have skyrocketed.

TRENDS IN CONSTRUCTION 17.1, 17.2

- 3D Printed Buildings/ Structures
- Self-Healing Building Components
- Smart Components
- Weather Controlled Domed Cities
- Smart Cities
- Vertical Interconnected Self Sustained Communities
- Translucent Concrete
- Building Information Modeling – BIM
- Site Automation/ factory automation

Promising Tech in Construction/housing

Other trends in housing include:
- Green buildings/ Smart buildings
- Rammed earth houses
- Cob houses
- Houses with cement stabilized mud/earth blocks
- Earthship/ earthbag houses
- Green roofing/ turf roofing
- 3D printed houses
- Prefabricated houses
- Modular houses
- Sustainable Housing Communities - Zero Carbon/ Carbon Neutral Technology for Water, Power, Food, Waste Management.

List for further exploration:

_____

_____

𝒥𝓃𝓈𝓅𝒾𝓇𝒾𝓃𝑔 𝒥𝓃𝓋𝑒𝓃𝓉𝑜𝓇𝓈

The labourer who studied chemistry and improved cement

Isaac Charles Johnson (1811-1911) is the inventor of true Portland cement. From the age of 16 while studying chemistry, he worked as a laborer at a "Roman Cement" plant, where his father was a charge-hand. After two years of persistent hard work, he succeeded in creating cement superior and cheaper than Joseph Aspidin's Portland cement. He ultimately took over Joseph Aspidin's cement plant as he was driven out of business by Johnson's superior product. Johnson pioneered several innovations including low-water raw mix slurries, improved kilns, and chimneys. On his 100th birthday, representatives of the cement industry presented Johnson with a silver tea service [17.3, 17.4].

STEP 18

HEALTH - GENETICS - FOOD

Of all the forms of inequality, injustice in healthcare is the most shocking and inhumane -
Martin Luther King, Jr.

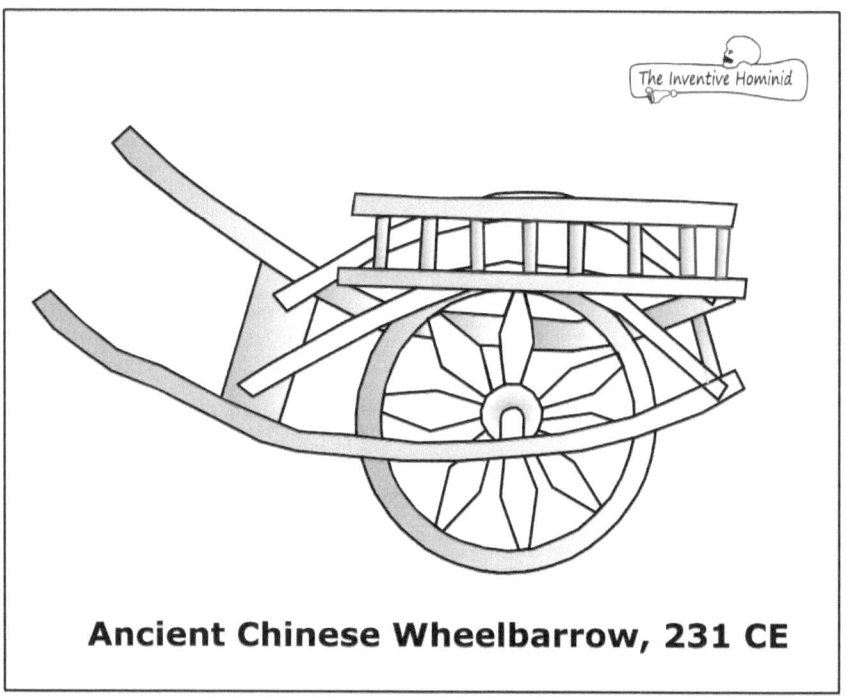

The Inventive Hominid

Ancient Chinese Wheelbarrow, 231 CE

TIH - Wheelbarrow (18 IR 1,18 IR 2,18 TR 1)

Learning Objectives

After studying the contents of this chapter, you should be able to:
- Learn about emerging trends in healthcare and food systems
- Learn about the increasing role of genetic engineering
- Understand the role of GMO

Biology – can we ignore something as basic as Life? Bioengineering, agriculture, Genetics, call it by any name, you just can ignore the importance of Life and life sustaining systems. Hence, areas such as Health Care and Agriculture are undergoing a profound shift as we advance our understanding in the field of biochemistry that governs our genes and processes inside us.

TRENDS IN HEALTH CARE 18.1

o Liquid Biopsies for Cancer / Disease Detection
o Electroceuticals
o Plasmonic materials
o Gene Drive
o Human Cell Atlas - To Identify Type and Function of Every Type of Cell in Every Tissue
o AI led molecular design
o Genomic Vaccines - Vaccines Based On Genes
o Implantable drug making chips
o Organs on Chips - Miniature Models of Human Organs
o Perovskite Solar Cells
o Optogenetics for Brain Disorder Treatment
o Systems, Metabolic Engineering - Synthetic Biology, Systems Biology, Evolutionary Engineering
o IOT Based Wearables for Disease Diagnosis and Treatment Monitoring
o Real Time Health Care - Sensors/ Wearables/ GPS/ GIS Based Diagnosis and Treatment
o 3D Printed Drugs, Artificial Uterus
o Bioelectronics - Miniaturized Implantable Devices

Heard of Tomatoes growing on a potato plant?

Biological Engineering or Bioengineering is a discipline founded upon the biological sciences. When Electrical engineers began to develop medical devices, they struggled with their poor knowledge of biological processes. Hence, bio-engineers devoted more time to study bioprocesses to develop better medical devices. The first Bioengineering program was created at University of California, San Diego in 1966 [18.2, 18.3].

Due to the complexity of life and associated bioprocesses, the field of bioengineering encompasses a wide range of sub-fields such as:

Tissue Engineering, Neural Engineering, Bioinformatics, Genetic Engineering, Biomechanics, Biomimetics, Bionics, etc.

Yes, you guessed it right; such plants are called Pomato!

EMERGING TECH IN FOOD SYSTEMS (AGRICULTURE, LIVESTOCK) 18.2

There are people in the world so hungry, that God cannot appear to them except in the form of bread. -Mahatma Gandhi, The Spirituality of Bread

- o Advanced Air, Soil, and Crop Sensors
- o Livestock Biometrics
- o Genetically Designed Food
- o In-Vitro Meat
- o Agri-Bots
- o Drone Based Crop Monitoring and Production
- o Vertical Farming Systems
- o Climate Resilient Farming
- o Closed Loop Farming Systems
- o Precision Farming - With GPS, Sensors, GIS, Data Analytics, Drones
- o Water from Air Tech
- o Insect farming for sustainable protein

The ultimate goal of farming is not the growing of crops, but the cultivation and perfection of human beings. - Masanobu Fukuoka, The One-Straw Revolution

List for further exploration:

_____

_____

Genetically Modified Organisms (GMO) – Saviour or Villain

As the population of humans in our world is expected to hit 9.1 billion (910 crores), food production, it is estimated must increase by 70% between 2005 and 2050 to feed this huge numbers. China's population will fall from the current 1.39 billion to 1.34 billion in 2050 and India's population will rise from the current 1.37 billion to 1.68 billion in 2050 [18.4].

GMOs are seen as a promising technology to feed this growing numbers. What is a GMO? GMO is any organism whose genetic material has been altered using genetic engineering techniques [18.5]. Humans have been selectively breeding plants and animals for specific genetic traits since around 12,000 BCE. Herbert Boyer and Stanley Cohen created the first GMO in 1973, when they took a gene from one bacterium and inserted it into another bacterium through a plasmid [18.6]. Since then scientists have created GMOs out of every conceivable microbe, plant, animal, and fish.

It is possible that the food we eat today has a fair amount of GMO in it and the safety concerns raised by one section of scientists have started a long running conflict between pro-GMO and anti-GMO groups worldwide.

Research into creating safe GMO food crops is an exciting career area for those with an interest into bioengineering.

STEP 19

ENERGY

We must get rid of fossil fuels by developing injection systems for automobiles, which can run on bio-fuel. - A. P. J. Abdul Kalam

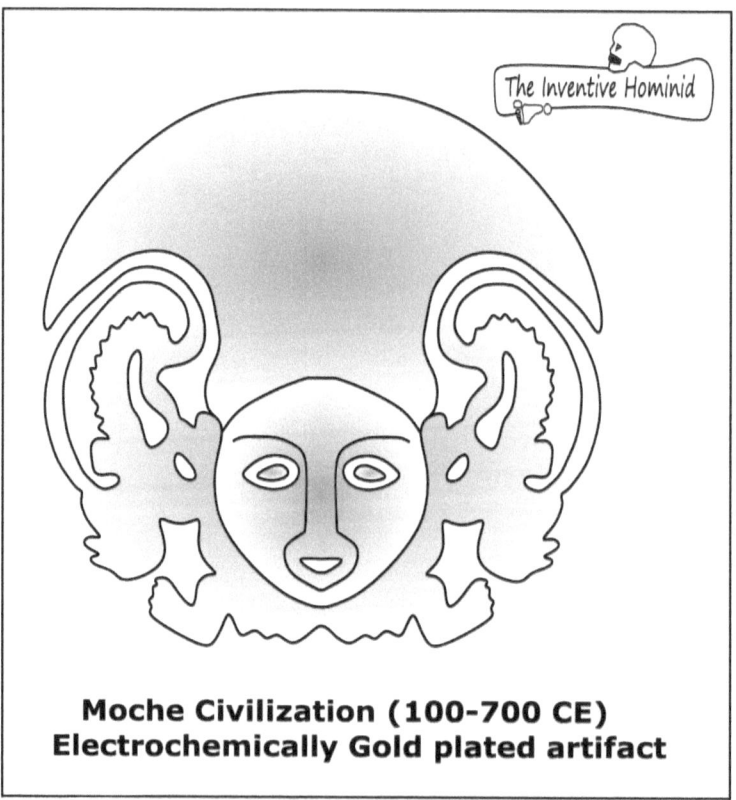

Moche Civilization (100-700 CE)
Electrochemically Gold plated artifact

TIH - Moche Gold Artifact (19 IR 1, 19 TR 1)

Learning Objectives

After studying the contents of this chapter, you should be able to:
- Learn about the emerging technologies in energy sector
- Learn about the link between fossil fuel use and climate change
- Understand the need for renewable energy technologies

The importance of energy in human advancement can never be understated. In our insatiable quest for energy, we have pushed our planet into uncharted territory of global warming. Hence, it is an urgent need that we reduce our CO_2 emissions while finding alternatives to satisfy our energy needs. Do you want to join the noble quest of deriving our entire energy needs from renewable sources? Go ahead and join the glorious team of our only star the Sun.

EMERGING TRENDS IN ENERGY TECH 18.2

- Lithium Air Batteries
- Hydrogen Energy
- Airborne Turbines
- Artificial Photosynthesis
- Home Fuel Cell
- Nanowire Battery
- Wireless Energy Transfer
- Passive Energy Buildings
- Thermal Storage in Insulated Repositories
- Smart Energy Network
- Micro/ Nano Engines
- Smart Solar Panels
- Integrated Solar Cells
- Advanced bio-fuels
- Space Based Solar Systems
- Ultra Efficient Batteries
- Liquid Fuel from Solar Radiation
- Zero Emission Fuel Cells for Vehicles

List for further exploration:

_____

_____

Inventions: From Whale oil to Petroleum

Humans have relied on burning wood from time immemorial to fuel their energy needs (cooking and heating). Usage of coal from the 16th century CE averted a disaster fueled by widespread wood shortage resulting from deforestation. Whale oil was used widely in lamps and soaps. As a result, whales were almost hunted to extinction around the world. The hunting of whales for oil declined with the use of kerosene for energy needs [19.1].

The usage of petrol and diesel in the industries and automobiles accelerated industrial development from the 1900s. With much of the advancement in our lives coming from petroleum and its derivatives which range from plastics to polymers.

Acceleration of industrialization among the developing countries has increased the emission of CO_2 pushing our climate balance to the brink of unpredictable and uninhabitable proportions. Hence the urgency to switch over to clean sustainable energy.

While ancient humans consumed enormous amounts of food to fuel their lifestyle of hard manual labor, we modern humans consume even higher quantities of food, but without expending it due to our sedentary lifestyles, leading to obesity across the global population.

STEP 20
MOBILITY

Once there was an old man who lived in a tiny village. Although poor, he was envied by all, for he owned a beautiful white horse -Max Lucado

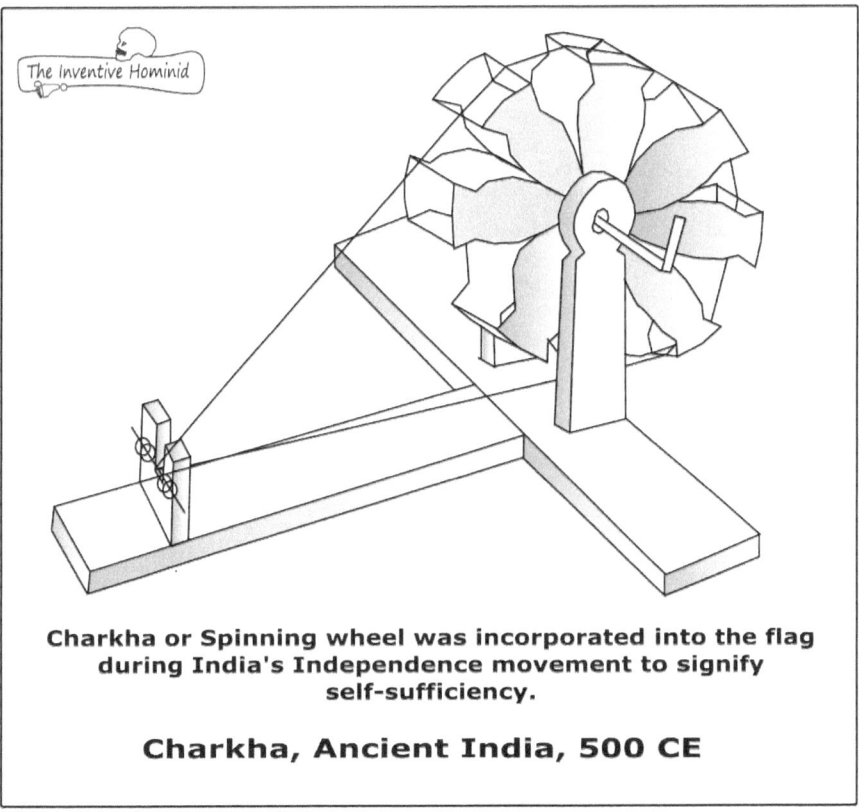

Charkha or Spinning wheel was incorporated into the flag during India's Independence movement to signify self-sufficiency.

Charkha, Ancient India, 500 CE

TIH - Charkha - Spinning Wheel (20 IR 1, 20 TR 1)

Learning Objectives

After studying the contents of this chapter, you should be able to:
- Learn about the role of mobility in our lifestyles
- Learn about the emerging technologies in the mobility sector
- List areas of your interest for further study

Efficient mobility has long fascinated humans as we marveled at the ability of birds to traverse thousands of kilometers/ miles to their wintering grounds.

We have been titillated at the idea of teleportation in sci-fi movies over the decades. Recent concerns over the impact of fossil fuels on our climate have spawned research into clean mobility. Hence, the current trend ranges from electric vehicles, driverless vehicles, vacuum tube transport, human quadcopters, bullet trains, bicycle sharing/ hubs, etc.

With the advent of ICT, mobility is an area, which is seeing a convergence of technologies like GIS, electric storage devices, driverless navigation, etc. it is an exciting, and evergreen area to get into and show your mettle.

FUTURE OF MOBILITY 20.1, 20.2

Autonomous Vehicles/ Machines/ Armaments,
Airless Tires,
Advanced Fuel Cells,
Beam Powered Propulsion,
Hyperloop,
Flexible Wings,
Flying Car,
Hover Vehicles,
Jet Packs,
Nuclear Rockets,
Photon Rockets,
Reusable Launch Systems,
Space Elevators.

List for further exploration:

_____

_____

Inspiring Inventors

The automobile revolutionary who worked for Edison

Henry Ford (1863-1947) founder of the Ford Motor Company was born in his Family Farm in Michigan, USA. He was interested in mechanical devices from an early age gaining a reputation as a watch repair technician. Not interested in farming, he left home at 16 to work as an apprentice machinist in Detroit. Through hard work, he rose to become the Chief Engineer in 1893 at Edison Illuminating Company of Thomas Alva Edison. In his spare time, Ford developed an automobile named Ford Quadricycle in 1896. After two failed attempts to manufacture a car, in his third venture with Alexander Malcomson, after early financial troubles, introduced the iconic Ford Model T in 1908 [20.3] . This became a huge success and sold 1,500,734 cars until 1927. Highly efficient production at Ford was due to the introduction of the pioneering assembly line technique of manufacturing introduced in 1913, which today is followed in every factory in the world.

Henry Ford was a pioneer who introduced the following worker welfare measures: doubled the wages to $5 per day, introduced the 5-day workweek, and offered profit sharing to workers who did not drink or gamble. He collaborated with Edison to research into many emerging technologies. Ford carried on his spirit of invention well into old age and was awarded 161 US patents [20.4, 20.5].

STEP 21

LONGEVITY

A good man never dies. - Callimachus, Epigrams, X

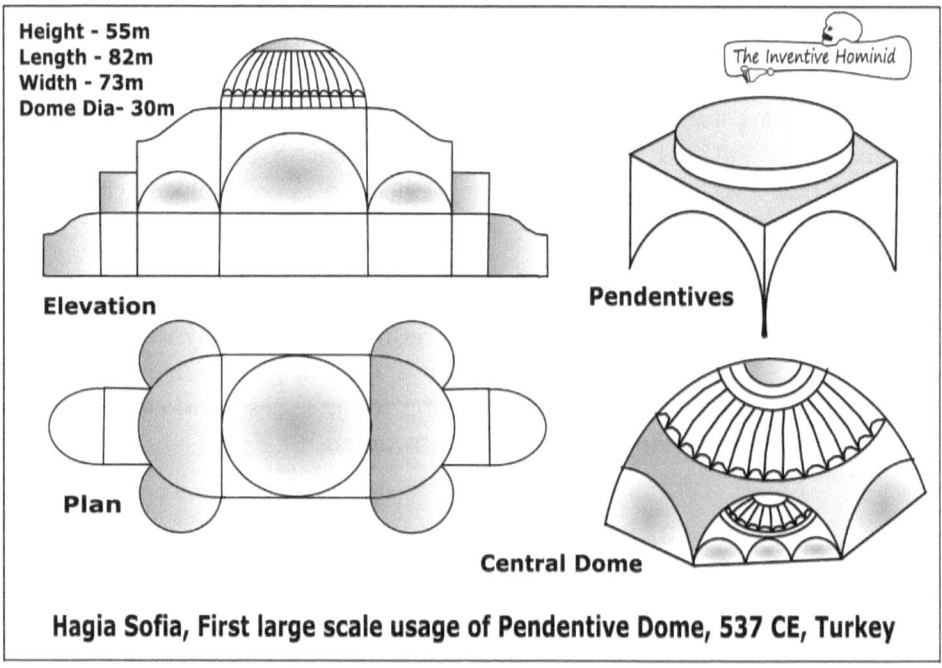

Height - 55m
Length - 82m
Width - 73m
Dome Dia- 30m

The Inventive Hominid

Elevation

Pendentives

Plan

Central Dome

Hagia Sofia, First large scale usage of Pendentive Dome, 537 CE, Turkey

TIH - Hagia Sophia - Pendentive Dome (21 IR 1, 21 IR 2, 21 TR 1)

Learning Objectives

After studying the contents of this chapter, you should be able to:
- Learn about the efforts underway to extend the life of humans
- List areas which you would like to explore further in longevity

Humans since ancient times have been trying to figure out a way to extend our life span. Various treatments to delay the onset of degeneration of skin and hair have existed for centuries and today they are a multi-billion dollar industry.

Recent times have spawned companies, which offer to preserve your body in a super cooled liquid after your death, for revival into life in the future when science advances to that stage. Companies are offering to preserve your brain and download its contents (consciousness) into a storage device to be uploaded later into some other body and brain.

Research has picked up in the area of life extension science and is focusing on areas such as genome editing, telomerase treatment, young blood transfusion, cranial stimulation, gene fooling, cryogenic preservation, Xeno-transplantation, anti-aging drugs, etc.

For more information, you can look up the resources on Life Extension in Wikiversity [21.1].

EMERGING AREAS IN LONGEVITY RESEARCH 21.2

- Head Transplants
- Body/ Brain Implants
- Cryonic Life Extension
- De-Extinction
- Induced Hibernation
- Synthetic Hormones
- Young Blood Anti-Aging
- Gene Editing and Splicing

List for further exploration:

_____

_____

Meet the tree which was born during the reign of the Pharaohs

A 5000-year-old tree belonging to the species Pinus Longaeva (Bristlle cone pine) is currently the oldest known living tree [21.3].

A quahog clam (Artica Islandica) was recorded to have lived unto the age of 507 years.

One bowhead whale killed in a hunt was recorded to be around 245 years old [21.4].

Adwaita was a male Giant Aldabra Tortoise caught from Aldabra Atoll in Seychelles, and given to Major General Robert Clive of East India Company in Kolkata. Transferred to the Alipore Zoo in 1875, it died in 2006 at an age of 250 years [21.5].

STEP 22

OTHER EMERGING AREAS

The future cannot be predicted, but futures can be invented.
-Dennis Gabor, Inventing the Future (1963)

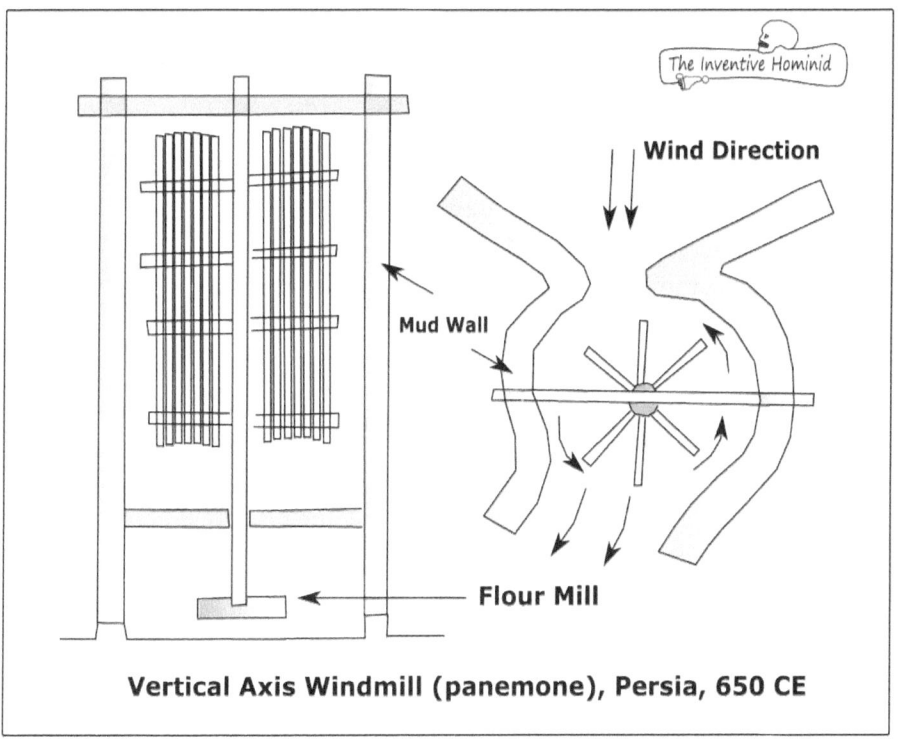

Vertical Axis Windmill (panemone), Persia, 650 CE

TIH - Windmill (22 IR 1, 22 TR 1)

Learning Objectives

After studying the contents of this chapter, you should be able to:
- Understand the emerging areas in which the world of innovation is focusing
- List areas which you would like to combine and come up with a new technology

It is impossible to list and dwell into all the areas our engineers and scientists are working currently to push the frontiers of technology.

Hence, let us look at a short list of areas and technologies, currently being investigated [22.1, 22.2, 22.3].

AVIATION

- Nano-Micro-Aero Bots/ UAVs
- Biomimetics Based Aerial Devices
- Neural Sensing/ Controlling Headgear - Brain-Computer Interface
- Smart skin
- Non-metallic gas turbines

MATERIAL SCIENCE

- Aerogel
- Bioplastics
- Metamaterials
- Nanomaterials
- Programmable Matter
- Quantum Dots
- Superalloy
- Synthetic Diamonds
- Time Crystals
- Nanoclay polymer composites

DISPLAYS

- 3D Displays
- Ferroelectric LED
- Holography
- Laser Video Displays

o   Screenless Display (Retinal, Bionic Contact Lens, Eyetap, Googleglass).

Inspiring Inventors

LED market is worth $50 Billion, but its inventor died of starvation

Oleg Vladimirovich Losev (1903-1942) was a Russian scientist and inventor credited with discovering practical application of LED [22.4, 22.5]. Although he never received college education, Losev conducted some of the pioneering research into semi-conductors and published 43 papers and received 16 patents [22.6].

As Losev was born in a noble family, he was denied college education after the Russian revolution. He was never promoted beyond the level of a technician. Throughout his career, he continued his research without any encouragement or support. He researched into LED and proposed the correct theory of how they worked and suggested practical applications. He researched into semiconductor junctions, built first solid-state amplifiers, electronic oscillators, and superheterodyne radio receivers, well before the invention of transistors.

He was finally awarded a PhD for his work in 1938, very late in his career. Losev died of starvation at the age of 38 along with many civilians, during the Siege of Leningrad in World War II. Only recently was Losev given credit for his path breaking inventions under extreme hardship.

ELECTRONICS

o   Biomimetics
o   Biometrics
o   Electronic Nose
o   Flexible Electronics
o   Spintronics
o   Nanoelectronic Machines
o   3D Circuits
o   Thin-film ceramic sensors

DEFENCE/ MILITARY

o   Caseless Ammunition
o   Camouflaging Devices
o   Electrolaser

- o Forcefield
- o Green Bullet
- o Plasma weapon
- o Sonic Weapon

SPACE

- o Anti-Gravity
- o Artificial Gravity
- o Asteroid Mining
- o Hypertelescope
- o Space Habitats
- o Micro/ Nano Satellites
- o Self Contained Habitats

ROBOTICS

- o Humanoids
- o Transformer Robots
- o Swarm Robotics
- o Powered Exoskeleton
- o Smart UAVs

CONVERGENCE OF NANO-BIO-INFO-COGNO 22.7

Just imagine the possibilities of convergence of all these emerging technologies: Nanotech, InfoTech, Biotech, and Cognitive Tech. It heralds infinite possibilities in the way we live, eat, entertain, heal sickness, and extend our life span.

Join the Trans-humanists in advancing the frontiers of convergence technologies. Form a convergence club in your Class or department or Institute, brainstorm the possibilities, and immerse yourself in this wonderful intellectual pursuit.

List ideas for combining various technologies into a product:

_____

_____

Inspiring Inventors

Inspiring Inventions: Space is the final frontier

The Space beyond our Earth has long fascinated humans. The perceived motion of the sun, the moon, and their eclipses, planets and comets filled humans with awe and dread. They were believed to control our destinies and human sacrifices were made to appease them.

Since the advent of science and exploration of planets through telescopes, orbiters and landers, our curiosity has only been partially satisfied. The efforts to find an inhabitable planet like our mother earth have spawned a race into space exploration in the recent years.

We have landed rovers for the first time on the moon (1970-USSR), Venus (1970-USSR), Mars (1971-USSR), Asteroid 433 Eros (2001-USA), and Titan (2005-ESA-USA-Italy). We have sent spacecrafts on flybys to the moon(1959-USSR), Venus (1961-USSR), Mars (1962-USSR), Jupiter (1973-USA), Mercury (1974-USA), Saturn (1979-USA),Titan (1980-USA), Uranus (1986-USA), Neptune (1989-USA), Asteroid 951 Gaspra (1991-USA), Ceres (2015-USA), Pluto (2015-USA) [22.8, 22.9].

There have been numerous other orbiters and rovers, which have sent astonishing photos and data regarding the terrain, soil, atmosphere of distant planets and asteroids. Stephen hawking has predicted that humans will go extinct if we do not find another planet to live within the next 1000 years. It is an exciting time to be an Aerospace engineer.

The Indian planetary program reached its zenith when it put its Mars Mission Orbiter on mars orbit in 2014, to become the least expensive interplanetary mission ever [22.10].

STEP 23

REVIEW THE PROBLEMS IDENTIFIED

The main source of problems is solutions - Eric Sevareid.

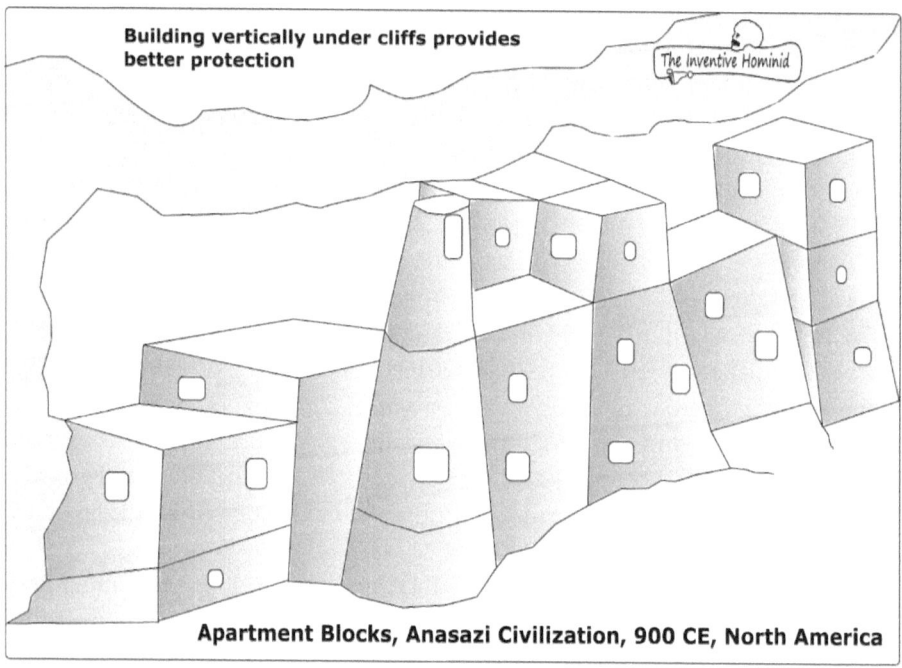

Building vertically under cliffs provides better protection

The Inventive Hominid

Apartment Blocks, Anasazi Civilization, 900 CE, North America

TIH - Apartments - Anasazi (23 IR 1, 23 TR 1)

Learning Objectives

After studying the contents of this chapter, you should be able to:
- Learn about reviewing the identified problem against various criteria
- Understand the importance of various criteria to check the relevance of identified problem

REVIEW YOUR PROBLEM

Once you have made up your mind regarding the problem you are going to solve, you are just a step away from diving into the problem. However, before that, check whether the problem you have selected ticks the following boxes?

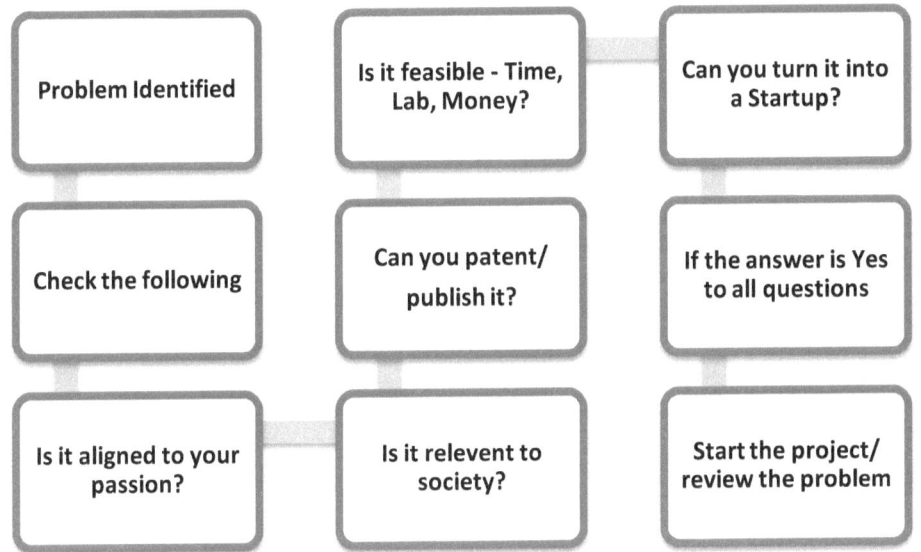

IS IT ALIGNED WITH YOUR CAREER/ PASSION?

Is the problem aligned with your passion? Do not choose a problem that is not in your passion zone in terms of subjects/courses or concepts. A big mistake people make is to pick a problem based on its popularity or hype in the media. Always choose a problem inside your passion zone. Your passion will ensure that you will overcome any obstacle that you might encounter. However, if you are persistent enough and have the will power, you can pick any problem and start from scratch.

CHECK FOR RELEVANCE

Is your problem going to solve a very critical problem in the society? Check whether your problem will solve a big enough problem in the society. Remember, we are not here for scraps; we are after bigger fish suitable for our high abilities.

CHECK FOR FEASIBILITY

Labs, expenses, time, support staff, etc.

Make a list of your needs for the project: the equipment you are going to need to run your experiments, the raw materials, and components you need for the project. Are these readily available where you are based? Can you access the labs that you need for your project? Can you accomplish your objectives within the available time?

Of course, if you have a sure shot idea that can make a lot of money, you need not worry about resources, the right people will be ready to offer the required resources.

CAN YOU PATENT AND COMMERCIALIZE IT?

Once you are ready with your results, are the final product / process / design patentable? If you are looking for commercial success, then being able to patent your final output is a critical step

To know details about patents and patenting process, kindly refer chapter no. 29.

Inspiring Inventors

The hungry penniless man who established the World's First Industrial Research laboratory

Thomas Alva Edison landed in New York from Boston in 1969, feeling very hungry. Not having money for breakfast, he went into a tea-grading house and asked if he could have some. After drinking some tea, he walked over many miles to meet his friend. His friend who was without a job at that time could lend him only one dollar, which was enough money for bed and breakfast for a few days. Unable to find a job, he stayed in the battery room of the Gold Indicator Company. On the third day of his stay, there was commotion due to a technical problem. Edison stepped inside, solved the problem, and was appointed as In-charge of the whole plant for a salary of $300 per month [23.1]. Edison improved the Gold indicators and stock tickers of his company. When his boss asked to sell all his patents to him, he thought of selling it for around $5,000 or even $3000. However, unable to ask for such a huge sum of money, he instead asked his boss to quote a fair price. Edison was astonished when the President of the company offered him $40,000. When Edison cashed the check, people at the bank played a joke on him, by offering his money in small bills put together in a big sack. Edison carried it around all night and did not sleep with so much money lying around. Next morning, the bankers having had enough of their joke, opened a bank account for him and deposited the money in it [23.1].

With this money, he established his workshop and began his torrent of inventions. In his Workshop, the first organized industrial research lab in the world, Edison stocked almost every conceivable material. The lab contained 8000 chemicals, all types of screws, all sizes of needles, all types of cord or wire, hair of humans, hogs, cows, rabbits, camels, mink, goats, silk of all textures, cocoons, various kind of hooves, shark's teeth, horns of deer, shell of tortoises, cork, resin, varnish, oil, ostrich feathers, tail feathers of peacocks, amber, all ores, etc [23.2].

STEP 24

Problem Selected, Reviewed and Confirmed!

"I've always thought that schoolchildren should be marked by the number of failures they've had. The child who tries strange things and experiences lots of failures to get there is probably more creative."
– James Dyson, Inventor of the bagless vacuum cleaner.

The library at Nalanda consisted of 3 multistories buidings housing hundreds of thousands of manuscripts, with Ratnodadhi (Sea of Jewels) being 9 stories high.

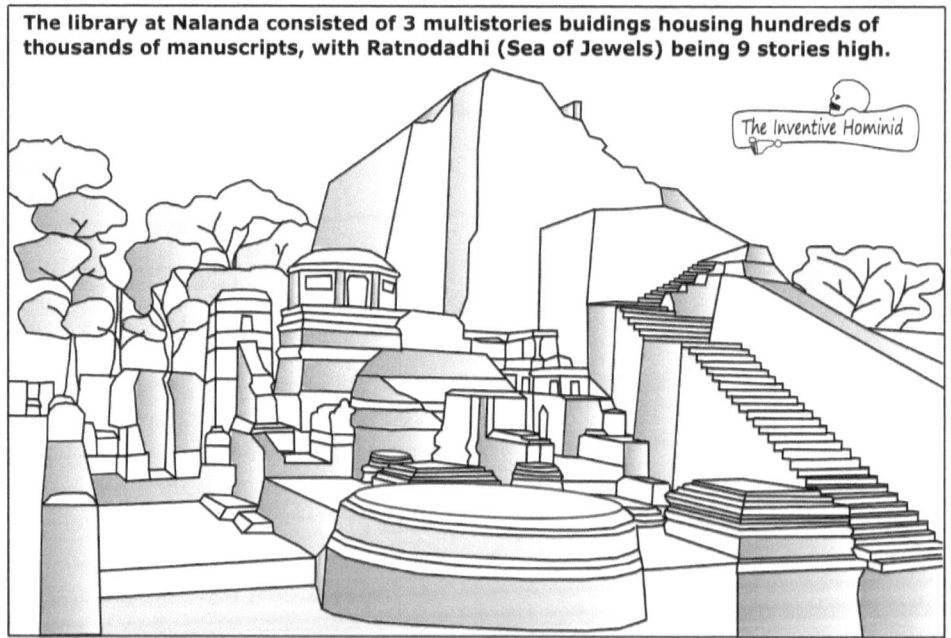

Nalanda University, 5 - 1200 CE, India. Was a centre of learning attracting students from Tibet, China, Korea and Central Asia. It housed around 2000 teachers and 10,000 Students. Its curriculum included Buddhist Philosophy, Vedas, Logic, Grammar & Philology, Magic and Samkhya.

TIH - Nalanda University (24 IR1, 24 TR1)

Learning Objectives

After studying the contents of this chapter, you should be able to:
- Understand the questions that need to be answered before embarking on your world-changing project

Good, now you have reviewed and confirmed your problem.

HOW TO START SOLVING THE PROBLEM?

Let us imagine, you have decided to build an astonishing humanoid robot! Well then, jump into the topic.

Step 1: scour your library and get all the books directly or remotely connected to robotics.

Step 2: easy: read all the above books. Read, read, and read again until you understand the concepts of building a robot. Stuck at some place, do not worry and get online.

Ask the questions to everyone and everywhere on the internet in the hundreds of robotic forums. Rest assured you would get tons of encouraging answers and directions for further reading.

After understanding the latest knowledge, now design the system, build a prototype, test, and produce the final version. Simple, is it not?

Oops, more reading? Yes, remember, we are in a marathon and we are NOT going to give up! Throw the entire library and I am going to gobble them all! This should be your attitude!

How should you go about to know the "Current state-of-the-art" (CUSA) (where the knowledge stands) in any topic?

Moreover, what is a project supposed to deliver? What are the challenges we might encounter in managing a project? Let us explore these questions.

Inspiring Inventors

The man who kick-started the cotton revolution 24.1

Eli Whitney (1765 – 1825) was an American inventor known for the Cotton Gin (machine that removes seeds from cotton fibers). He was born to a prosperous farmer and lost his mother when he was 11. Unable to attend college due to his stepmother's objections, Whitney worked as a farm laborer and schoolteacher for a number of years to save money for college. He attended college at 23 and graduated from Yale with top honors in 1792. Not having money to study law, he accepted an offer to become a private tutor.

On an invitation from an acquaintance he met on the ship, he travelled to a Georgian farm. There, upon the request of cotton growers for a machine to remove seeds from cotton, he created the Cotton Gin. Able to process more cotton, the Gin made growing cotton profitable and cotton exports from US to Europe exploded in the following years. Whitney hoped to build and rent the machines himself, but the demand was so great that many copied his machines and flooded the market. Whitney spent a major portion of his earnings to sue the counterfeit machine makers, but could not profit much from the Gin.

Whitney then turned his attention to manufacture muskets with interchangeable parts and won a huge government contract. He remained a man creating many inventions until his death.

STEP 25

PLAN & MANAGE YOUR PROJECT MEN, MATERIALS, MONEY, MILESTONES

In the past few years, it has been recognized that project management is an identifiable profession — or job category, if you like — requiring specific skills, training, knowledge and even cultivated personality traits -Michael Marcus

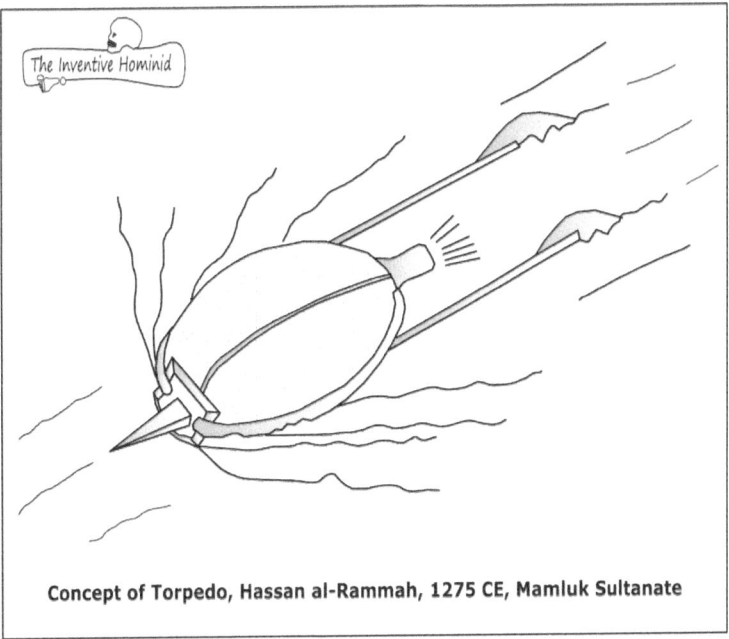

The Inventive Hominid

Concept of Torpedo, Hassan al-Rammah, 1275 CE, Mamluk Sultanate

TIH - Torpedo (25 IR 1, 25 TR 1)

Learning Objectives

After studying the contents of this chapter, you should be able to:
- Understand the various elements that need to be successfully managed in a project

The skill of managing a project is a much sought after one in today's world. It is a cross functional skill, which can be applied to any sector as the elements that constitute a project are the same.

WHAT IS A PROJECT?

As per the Project Management Institute (pmi.org), the leading Project Management Certification Organization in the world [25.1],

"It's a temporary endeavor undertaken to create a unique product, service, or result." It also defines Project management as "the application of knowledge, skills, tools, and techniques to project activities to meet the project requirements."

PMI's Guide explains that effective Project management requires knowledge on ten areas [25.2].

TEN AREAS OF PROJECT MANAGEMENT AS PER PMI

- o Integration
- o Scope
- o Time
- o Cost
- o Quality
- o Procurement
- o Human resources
- o Communications
- o Risk management
- o Stakeholder management

Any project can be successfully managed by controlling these elements. Effective project management is often the difference between a successful and a failed venture, ranging from a research project to a multi-million dollar startup. Budding engineers will do well by being acquainted with these elements through a course or a PMP (Project Management Professional) Certification [25.1].

Now coming to your project, let us discuss these 10 elements in relation to the project you are planning to execute.

INTEGRATION

Integration deals with combining different processes related to designing, planning, procurement, and completing the outcome in a smooth manner to complete the project. If you are working in a group with members working on different components, then integration of these components into to final whole becomes a critical element for the success of the project.

SCOPE

Scope deals with the boundaries of your project. You have to decide the scope of your project that can be achieved with the available time and resources. If your scope is limited, then you will not have a project that reflects your potential. On the other hand, if the scope is too large, then you will not be able to complete it.

TIME

Time refers to the duration given to complete your project. Typically, colleges provide a year or two semesters to complete a capstone or major engineering project. You have to plan your milestones (time of achieving major project goals) within this duration. Of course, if you start your project in your first year or second year of study, you can be more flexible with time.

Plan your project with a simple bar chart. For a typical project, involving creating prototype equipment, the bar chart is given below.

Sl. No.	Project component	Project duration (months)							
		1	2	3	4	5	6	7	8
1	Problem Identification								
2	Review of lit./ patent								
3	Design								
4	Procurement/ fabrication of components								
5	Preliminary Assembly & testing								
6	Final testing and data sheet								
7	Preparation of project report								

COST

Cost refers to expenditure you will have to incur in completing the project. Cost components include the cost of mechanical, electrical, electronic components like circuit boards, circuit components, wires, sensors, raw materials, chemicals, etc., which are required for doing your project. You should also include the cost of travelling if necessary to collect data, cost of preparing your report, cost of filing a patent if planned, etc. A typical expenditure table is given below.

Sl.	Component	Cost ($)
1	Mechanical components	2000
2	Printed circuit boards	1500
3	Sensors/ other instruments	1200
4	Chemicals/ raw materials	500
5	Travel costs	500
6	Conference/ competition registration	500
7	Patent filing for startups	800
8	Project report	500
	Total cost	**7500**

Once you have estimated your budget, next step is to look for funding. The options available are sharing the expenses among team members or applying for funding from your university or external sources. If you are planning to apply for funding, then you need to start early depending upon the time duration taken by funding agencies to arrive at a decision. You need to prepare a funding proposal with the help of a faculty supervisor. The aspects involved in preparing a funding proposal are discussed in detail in chapter - 40.

QUALITY

Refers to the quality standard you will be adhering to define what will be acceptable as a successful outcome of your project. For example, the incandescent lamp was invented much before the time of Edison, but the filament did not last long enough to be of any use. Edison invented a filament that lasted long enough to be useful for various applications. It also refers to the quality standards such as ISO and other national standards such as ASTM in US, IS Codes in India.

PROCUREMENT

Procurement refers to the purchase of components necessary for your project. Availability of required components of the desired quality and affordable price at the right time is important for completing your project. Some components can be purchased off the shelf, but some have to be ordered and will take weeks for the manufacturer to supply it to you. Therefore, you have to research well in advance the components you need and when and where it will be available. Contact all possible suppliers and get their quotation for the supply of items you need along with price, payment terms (when should it be paid? Before or after delivery; cash or card or online), time needed for delivery, specifications, etc. Lower price does not mean better, as you have to ensure that the specifications and quality is as required. Many funding agencies require that you obtain quotations from at least three suppliers, compare and then purchase the best in terms of quality and price. These documents are important at the time of a financial audit of the funds given to you and need to be preserved. A typical comparison sheet is given below for reference.

Sl.	Component	Companies / suppliers/ vendors					
		XYZ Ltd.		Tools Inc.		Amazone	
		Cost $	Time	Cost $	Time	Cost $	Time
1	Mechanical components	2000	1 week	2200	1 week	1800	1 week
2	Printed circuit boards	1500	1 week	1500	1 week	1500	1 week
3	Sensors/ instruments	1200	1 week	1100	2 weeks	900	2 weeks
4	Chemicals/ raw materials	500	1 week	500	1 week	500	1 week

HUMAN RESOURCES

Refers to the humans involved in the project. In this context, it refers to the team you are working with. A team with members working in harmony and having all the required skill set will be a successful one. If you are doing it alone, then it will encompass the persons who might help you in the lab, workshop, etc.

COMMUNICATIONS

Communications refers to the flow of information to all people involved. Here it might be the team members, the project advisor/supervisor, and the review committee. Smooth flow of communication is essential to avoid misunderstandings and potential delays in the project.

RISK MANAGEMENT

Managing of potential risks to the completion of the project is essential. By keeping proper control on time, expenses, resources and processes, we can minimize and eliminate the risks. Protocols such as bio-safety, animal, or human testing, ethical treatment of volunteers, handling, and disposal of toxic substances need to be followed and documented to avoid negative risks.

STAKEHOLDER MANAGEMENT

People with interest in your project are referred to as stakeholders. They may range from people/ company who have invested in the project, agency that has funded your project, the team members, etc. it is important that the expectations of all stakeholders be aligned towards the agreed final goals of the project.

Inspiring Inventors

For India's grassroots Innovators, Sheer necessity is the mother of their inventions

Annasaheb Udagavi (born 1935) is a farmer in Karnataka state known for his low-cost inventions in the field of irrigation and agro-equipment. Annasaheb struggled for decades with low water availability and tried different crops without much success. Finally, he decided to cultivate sugarcane, but he had to overcome the barrier of low water availability. Sugarcane requires a huge quantity of water under the traditional irrigation method. He overcame the hurdle by deciding to adopt sprinkler rain guns. However, prohibitive cost of rain-guns available in the market forced him to design a low cost model himself. Finally, he succeeded in creating a low-cost rain gun model named chandraprabha that is priced at $50 (INR 3500) per rain-gun and the installation cost of piping network as per his design costs around $200 (INR 15000) per acre [25.3]. He profitably cultivated sugarcane with the help of his rain-guns by saving water, providing manure also through the rain guns, which also washes down pests in the sugarcane due to its high-pressure jet. He has also created a multipurpose sugarcane planter cum manure applicator cum stubble shaver. He has been awarded twice by National Innovation Foundation [25.4] India in 2001 and 2007.

National Innovation Foundation in involved in scouting such innovations among rural grassroots innovators, documenting them, helping them obtain IPR and commercialize them. Until now, it has documented around 300,000 technological ideas, innovations, and traditional knowledge practices. To further its mission, NIF partners with a number of institutions such as Honey Bee Network (HBN)[25.5], Society of Research and Initiatives for sustainable Technologies and Institutions (SRISTI)[25.6],   Grassroots Innovation and Augmentation network (GIAN)[25.7], Govt. Research Labs, Indian Patent Office, law firms, NGOs, etc.

STEP 26

Catching Up With State Of Art (CUSA) - The Knowledge Update

The Master said, "Yu, shall I teach you what knowledge is?
When you know a thing, to hold that you know it; and when you do not know a thing, to
allow that you do not know it; — this is knowledge."
- Confucius in The Analects2:17, as translated by Arthur Waley

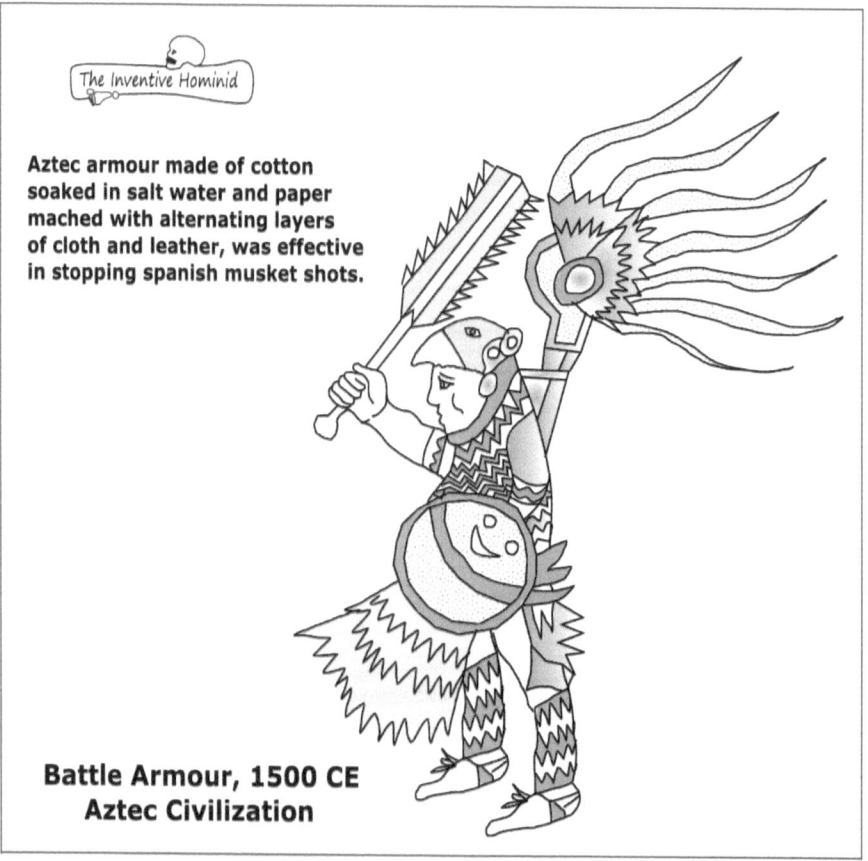

The Inventive Hominid

Aztec armour made of cotton
soaked in salt water and paper
mached with alternating layers
of cloth and leather, was effective
in stopping spanish musket shots.

Battle Armour, 1500 CE
Aztec Civilization

TIH - Aztec Armour (26 IR 1, 26 IR 2, 26 TR 1, 26 TR 2)

Learning Objectives

After studying the contents of this chapter, you should be able to:
- Learn about various avenues to update our knowledge
- Learn about online courses and video lectures that one can enroll freely

Catching Up With State of Art (CUSA)

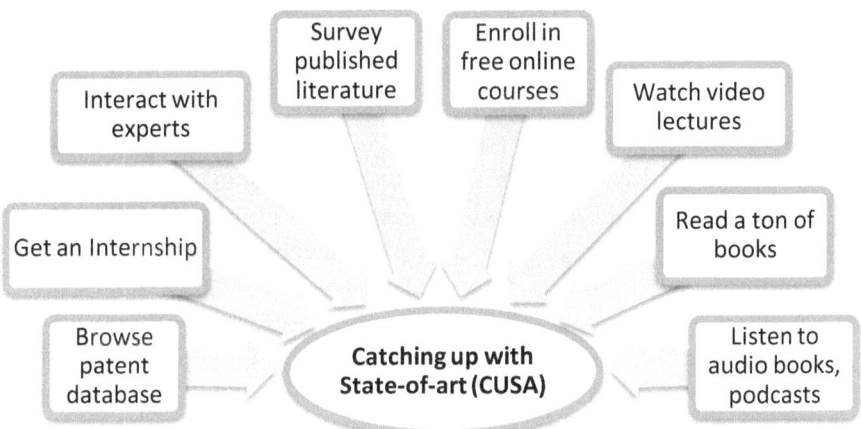

Enroll in a free online course

Go ahead and enroll for free online courses at edX (courses from top universities), Harvard, NPTEL, Coursera, etc. These sites offer courses on every topic under the sun: robotics, AI, plant genetics, environmental economics, you name it and they got it. If you are diligent with the course and follow the course instructions, you will be gaining a lot of quality knowledge without moving an inch right from your table.

Free video lectures

Do you feel like tuning in to the lecture of a famous professor at a top university? Then what are you waiting for, tune right in. Most online video sites have a big collection of video lectures by top professors. It is the best time to be an engineering student anywhere in the world, as they say, "geography has become history." All top universities are quite generous with their course lectures and many video course aggregation sites list subject wise lectures. Explore video lectures at MIT OCW, Khan Academy, Coursera, NPTEL, etc.

Inspiring Individuals

A free excellent education for anyone, anywhere.

This is the motto of Khan Academy, an e-learning educational organization founded by Salman Khan, an American born to a Bangladeshi Father and an Indian Mother. He is married to Pakistan-American Physician Umaima Marvi. He was awarded Padma Shri in 2016 by the Indian Government [26.1].

Salman Khan graduated with BS and MS degrees in Electrical Engineering and Computer Science and another Bachelor's degree in Mathematics in 1998, from MIT, USA. He also holds an MBA from Harvard Business School.

In 2003, when he started tutoring his cousin in mathematics over the internet, his friends and relatives also sought his tutoring. In response, he moved his tutorials to YouTube in 2006. Due to immense popularity of his YouTube videos, he quit his job and began focusing fulltime on his you tube channel, Khan Academy [26.2, 26.3].

As of today, Khan Academy in YouTube hosts 20,000+ videos in around a dozen of languages covering a wide range of topics including preparation videos for competitive exams. It has around 4.4 million subscribers learning every imaginable subject. It employs around 200 people and generated revenues of around $133 Million (2014).

STEP 27

CUSA – Get the Basics – Text Books

Some books are to be tasted, others to be swallowed, and some few to be chewed and digested - Francis Bacon, Essays (1625), "Of Studies."

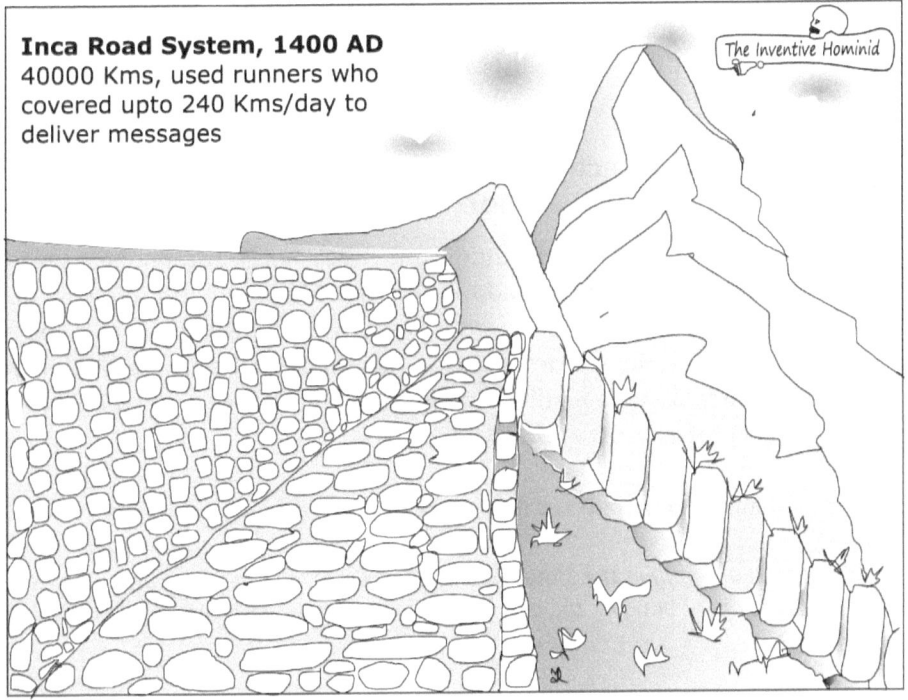

TIH - Inca Highway (27 IR 1, 27 TR 1)

Learning Objectives

After studying the contents of this chapter, you should be able to:
- Learn about exploring course material and book list of courses taught at top universities
- Understand the ways to concentrate fully when reading a book
- Know about audio books and its effectiveness in enhancing our knowledge

Reading list of the course at top universities

Log on to the course page on the subject of your interest and get the books and reading materials listed there. Having any difficulty? Write to the course Instructor requesting for the reading material and they will be happy to help

Try the authoritative reading lists in the courses/ subjects of MIT, Harvard, etc., just juice it, and drink it, no shortcuts.

Reading tricks and techniques: How to read a ton of books in a short time.

Focus on the reading task by cutting out distractions. The phone goes into silent mode (mute not vibration) and out of sight. Mute all notifications in your laptop/PC. Switch off that TV. Cover your windows to minimize glare. If you are in a noisy place, close the windows and switch on some steady music that you cannot understand. Go with meditation music or trance. This will cut off outside noise and let you focus.

o   Identify your concentration issues.

"Concentration signifies the state of being at a centre (con and centrum). Applied to thought, it is the act of bringing the mind to a single point" - Swami Mukerji, The Doctrine and Practice of Yoga.

Are you somebody who gets distracted even by the slightest noise? On the other hand, maybe you can read with full focus in the middle of noisy traffic. Are you an early riser or a night owl? Find out what you are comfortable with, and then schedule your reading time as per your circadian rhythms. Disrupting your natural sleep, wake cycle can affect your health.

- o Have a light that shines on the book from the top and not from the sides. Light should illuminate the book from the top without getting into your eyes. This will lessen eyestrain.
- o Start by meditating for a few minutes to focus your brain and shut off all residual thinking processes. At times, you will fall asleep doing this. That is fine, it means your brain requires some rest, sleep it off, and get up refreshed to start reading.
- o Use a pen or your finger to trace the lines you are reading. This will help you focus and increase your reading speed.
- o Do not stop to get the meaning of each word that you do not understand. At times, you can get the contextual meaning of a paragraph without actually understanding every single word of it.

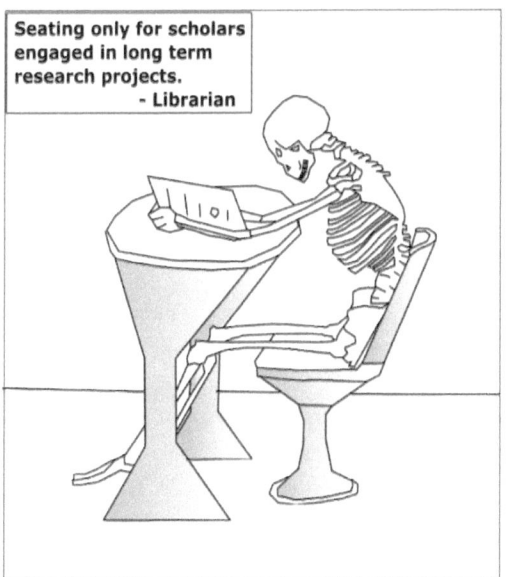

Seating only for scholars engaged in long term research projects.
- Librarian

Considering that, we are trying to understand technical stuff. It is better to focus on understanding the concept than speed. Once you are stuck with a concept, do not run around trying to find a person to clear your doubts. Get on the internet and type your question and you will find thousands of sources giving you the complete explanation with videos and slides in a second.

- o Have a notebook and take concise notes like a line or a diagram for each page or paragraph, you are reading. Write about what you have understood, doubts, and ideas. Use a pencil instead of a pen; it flows smoothly and is very good for drawing diagrams and corrections. You can also write notes in the margin of the book.
- o Feeling tired. May be your brain needs a refill of glucose or a bit of shuteye? If you hear growling noises from your tummy, then it wants food. You can never read productively on an empty stomach.

○   Never jolt your brain with excessive caffeine or sugar. After an initial burst of fireworks, your brain will get very tired. Go for a wholesome meal with less carbohydrates, more protein, and fiber.

○   Still feeling ragged after food and sleep, then what you need is a good workout. A good physical workout in fresh air will transport lots of oxygen to the brain and every tissue of your body and expel waste through sweat. A game or a jog/ walk in the fresh air are better than a workout at a closed-door gymnasium. The trick to think productively is by giving the right combo of food, sleep, and oxygen to your brain and not by jolting it with caffeine or sugar.

LISTEN TO THE BOOKS

Tired of reading? No problem. Just continue to listen to the books from your smart phone. Download your favourite audio books in mp3/audio format and listen to them while you commute or when you are tired to hold a book and engage your eyes. This way, even your English language pronunciation will improve from listening to a wide range of books.

Inspiring Inventors

The boy who read 2 books a day and is currently the Billionaire Tech Maverick of the world

Elon Reeve Musk (born 1971) is a tech entrepreneur and engineer. He is the founder of SpaceX; Boring Company, SolarCity, and co-founder of Tesla, Inc; Neuralink; Paypal and OpenAI. He is the 54th richest person in the world (2018 [27.1]).

While as a child, he taught himself programming at the age of 10. At 12, he sold the code of a video game he had created. He used to read for up to 10 hours a day and on weekends, he used to finish two books. He read all the books in his school library and the neighborhood library. When he ran out of books, he read the complete set of Encyclopedia Britannica [27.2].

He continues his reading habit even today and just read a number of books to become the founder and lead designer at SpaceX. This recently became the first company in the world to create reusable rockets, rockets that came back and landed on the launch pads. He is currently at the forefront of creating the most efficient electric cars with Tesla, Inc. We are sure to hear a lot about Musk and his path breaking innovations in the coming years.

STEP 28

CUSA - THE LITERATURE REVIEW

Research is something that everyone can do, and everyone ought to do. It is simply collecting information and thinking systematically about it - Raewyn Connell et al. (1975). [28.1]

The area covered by the Snowshoes should be large enough to spread the weight safely over the snow

The Inventive Hominid

Snowshoes, Alaskan Tribes, 3000 BCE

TIH - Snowshoes (28 IR 1, 28 TR 1)

Learning Objectives

After studying the contents of this chapter, you should be able to:
- Learn about the importance of literature review in a project

- Understand the process of accessing, downloading, reading journal papers from various databases

What is a literature review?

It is a meticulous survey of published research literature on a particular topic to understand the current research efforts, results, and persisting questions.

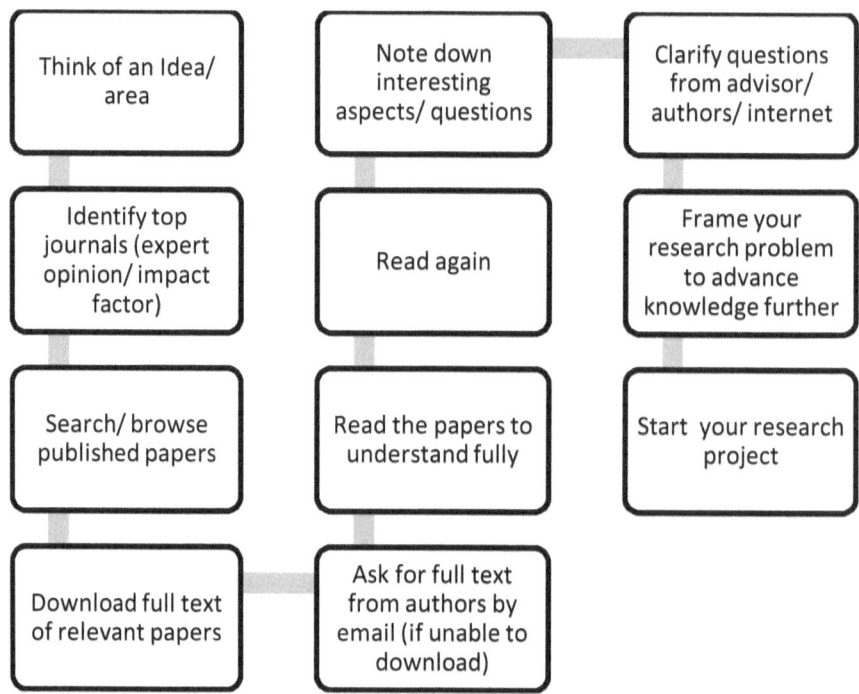

How it is done? In ancient times, before the advent of the internet, you were supposed to book your stay in the archives of your college library and take notes from each back volume of journals from the time it came into being. After some time people were able to take a photocopy of the papers of their interest. Still, they had to physically go through each volume of a journal and identify papers they want to photocopy. This was a huge time consuming exercise in the claustrophobic environs of archive rooms. Now the advent of internet has reduced this task to just typing a search term in the online databases.

There a number of research databases from which you can download the papers of your interest. However, choosing the right database is important, as the internet is full of papers from dubious sources. In this aspect, it is better to go by the advice of your faculty guide/supervisor. There are a number of research databases, of which some are subscription based and others offer free access.

Usually the library of an institution provides a link on their webpage to the databases, which have been subscribed by the institute. Then there are the open access publications, which offer free full text, download.

Some of the popular multidisciplinary research databases which are free include Google Scholar, African Journals Online (AJOL), BASE: Bielefeld Academic Search Engine, Directory of Open Access Journals, JURN, OAIster, etc.,

A meticulous literature review is half the job well done. Literature review gives you the information about what research has been done till now on the problem you have chosen; what are the results gained till now; what are the problems (research gaps) which still persist and which direction is the overall thinking headed.

With the above information, you will be able to unleash your thinking on what other directions you can move the research on. Choose a research gap as your research problem.

Inspiring Individuals

World's First Scientific Journal whose editor was arrested on suspicion of being a Spy

Philosophical Transactions of the Royal Society is the first scientific journal published in the world from 1665 [28.2]. The word Philosophical refers to natural philosophy, as science was referred to in early days. The responsibility of its publication including its financial affairs was of the Secretary of the Royal Society. The First issue was edited and published by Henry Oldebberg, the first Secretary of the Society in London on 6th March 1665. However, much to his disappointment it was not a financial success. In fact, it did not make any reasonable profit until 1948. The Society established rules for registration, peer review, dissemination, and archiving. Oldenberg was suspected of being a spy due to his frequent correspondence with foreign authors and was jailed in the Tower of London in 1667 for a brief period.

The Journal evolved over time to become a prestigious journal sought after by scientists to publish their papers. Notable scientists to publish their works include Isaac Newton, Benjamin Franklin, Charles Darwin, Michael Faraday, Alan Turing, Stephen Hawking, etc [28.2].

Since 1886, the journal was split into Philosophical transactions A and B, covering the Physical Sciences and Life Sciences respectively. Today everybody can access the issues of this journal right from 1665 freely available for download at http://rstl.royalsocietypublishing.org/.

STEP 29

CUSA - The Patent Review

A country without a patent office and good patent laws is just a crab and can't travel any way but sideways and backwards.
-Mark Twain, A Connecticut Yankee in King Arthur's Court (1889)

TIH - Native American - Aspirin (29 TR 1)

Learning Objectives

After studying the contents of this chapter, you should be able to:
- Learn about Intellectual property
- Patent classification as per US, Indian, and EU systems
- Understand the process of filing and obtaining patents for your inventions

You do the hard work for months or years together and come up with a wonderful piece of invention. Then, it turns out that your idea is similar to the one invented by Mr. Who-is-it in 1967. What? How can it be? Why me? No amount of crying will help the situation.

The trick is to avoid such a situation by doing a thorough patent search before you start your project.

Most Patent Offices offer a searchable database on their websites:

US Patent Office (uspto.gov) even offers a web-based tutorial for systematic patent search strategy. It provides a comprehensive searchable patent database for patents issued from 1976 and a patent number based database from 1790. Yeah, you read it right, it is 1790. It is important that you do not infringe on the Intellectual Property (IP) of other people/ company in your invention to avoid legal trouble.

What is Intellectual Property?

"Intellectual property (IP) refers to creations of the mind, such as inventions; literary and artistic works; designs; and symbols, names and images used in commerce"- World Intellectual Property Organization (WIPO).

WIPO further espouses the philosophy of striking a balance between protection and progress, "IP is protected in law by, for example, patents, copyright and trademarks, which enable people to earn recognition or financial benefit from what they invent or create. By striking the right balance between the interests of innovators and the wider public interest, the IP system aims to foster an environment in which creativity and innovation can flourish"- WIPO.

What is a patent?

A patent is a property right conferred for an invention to the inventor/s by the patent office of a country. This right excludes others from making use of the invention without the consent of the patent holder. Persons / companies interested in using the invention can buy the rights from the inventor.

Laszlo Biro came up with the idea of a ballpoint pen in 1944 and then sold the rights to manufacture and sell the ballpoint pens in North and Central America to Eversharp for 500,000 USD. However, Reynolds moved fast to manufacture and sell ballpoint pens before Eversharp by tweaking the ink flowing mechanism to sidestep patent infringement [29.1].

Many an inventor has earned nothing from their inventions, because of the fact that they did not obtain a patent for it. Some did not earn much even after patenting, because of the way they drafted the patent, which was exploited by competing companies to their advantage.

It is worthwhile to remember that patent will provide protection only in the country or region where you obtain a patent. Hence, depending on the commercial potential, a patent has to be obtained in multiple countries/jurisdictions.

The USPTO classifies patents as:
  o **Utility patent**

Utility patents may be granted to anyone who invents or discovers any new and useful process, machine, article of manufacture, or composition of matter, or any new and useful improvement thereof;

  o **Design patent**

Design patents may be granted to anyone who invents a new, original, and ornamental design for an article of manufacture; and

  o **Plant patent**

Plant patents may be granted to anyone who invents or discovers and asexually reproduces any distinct and new variety of plant."

However, the **EU patent office** does not include plant varieties as patentable; instead, it is governed by the Plant Variety Property Right and is granted a CPVR (Community Plant Variety Right).

The Indian Patent Office (ipindia.nic.in) provides patents for Inventions and registers Designs, Trademarks, and Geographic Indications.

The steps involved in obtaining a patent

o   Work on a unique idea
o   Prepare a sketch of your idea/ invention
o   Run a search on the patent database of the patent agency, you are planning to apply
o   Ensure that there is no patent on the idea, which you are working on
o   Understand the patent filing process in the patent agency, which you are planning to apply; all agencies have detailed presentations and guidance document explaining the whole process in their websites

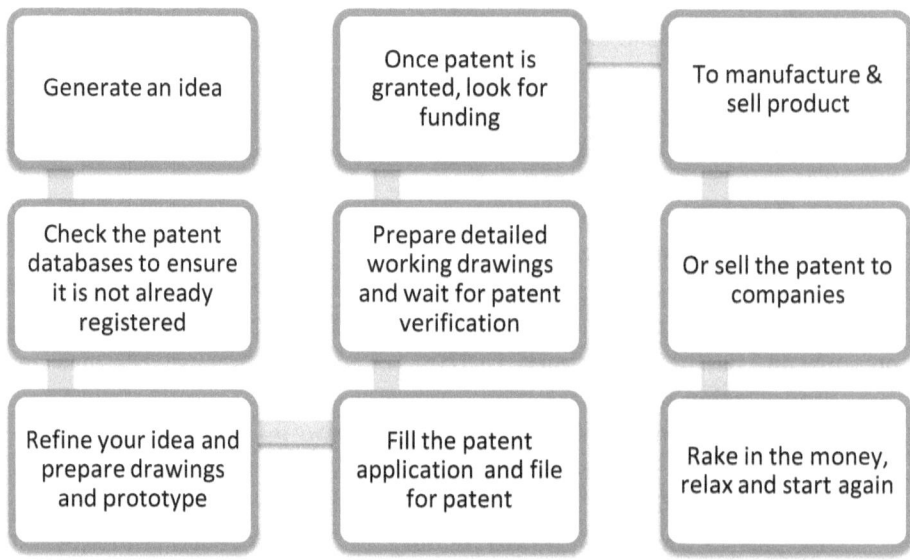

o   Prepare to apply for a patent by refining your invention and prepare a detailed description and drawings of your invention
o   Fill-up the patent application form yourself along with processing fees or take the help of patent consultants/ lawyers who will help but charge a fee for their services
o   File your patent application and wait for the verification process
o   A patent is granted after verification
o   Approach interested companies for commercialization through licensing or sale of patent rights
o   Rake in the money and relax!

Inspiring Inventions

The first patent system

The first patent system can traced to the Venetian Patent Statute of 1474. It accorded legal protection for inventive devices for a period of 10 years. As venetians travelled, the patent system slowly diffused across Europe [29.2].

Today, under the World Trade Organization's (WTO) TRIPS Agreement, patenting system should be instituted in all WTO member states in all field of technology and the term of protection should be a minimum of 20 years [29.3]. Companies spend millions of dollars every year to obtain patents for their inventions and improvements.

Certain companies like Coca Cola chose to maintain their formula as a trade secret instead of patenting it. Such trade secrets are protected by non-disclosure agreements and labor law to prevent it from being leaked. Thus unlike patents, which expire after a certain number of years, these trade secrets, remains a secret until it is made public.

Edison and Joseph Swan both independently came up with improved designs of the incandescent lamp in 1879. Though Edison obtained a patent in the US, he could not obtain one in UK as Joseph Swan had obtained a strong patent for his invention [29.4]. Hence, they both decided to form a joint company in UK, named the Edison & Swan United Electric Light Company. This company known as Ediswan, sold electric lamps made of cellulose filament invented by Swan, while the Edison Company sold lamps using the bamboo filament outside of Britain.

STEP 30

CUSA – Get an Internship at a Research Lab, Company, or Startup

A gram of experience is worth a ton of theory - Robert Gascoyne-Cecil

Hammocks - Carribean Islands

The cord tension transferring the weight should be less than the shear capacity of the poles

TIH - Hammock – Caribbean (30 TR 1)

Learning Objectives

After studying the contents of this chapter, you should be able to:
- Understand the preparation required to obtain an internship at various organizations
- Learn about websites offering assistance in obtaining an internship

Identify the best research labs, companies and startups, which are working in the area of your interest and work towards getting an internship there.

To get an internship, you have to demonstrate a good level of knowledge in the area they are working. Hence, update your knowledge to current state of the art and write to them early to get an idea of their requirements. Having a research publication that demonstrates meticulous research skill will be an advantage when applying for research labs. Whereas showing a good streak of creativity, multitasking, and a go get it attitude is valued highly at companies and startups.

Make a list of application deadlines for internship positions and start applying at least 6 months in advance. Highlight your passion for the particular area you are applying. Though most internship goes to the academically accomplished candidates, it is not the only criteria in many institutions. Internships in many situations lead to full time jobs or a good network, which can lead you to better jobs.

WEBSITES FOR INTERNSHIPS

Some of the websites for internships, which you can explore, are:
- Indeed
- Idealist
- Simplyhired
- LinkedIn
- Careershift
- Letsintern
- GoAbroad
- Experience
- MakeIntern
- Intern Desk
- Internshala

Inspiring Individuals

The man who did not join IIT after getting admission

Narayana Murthy (born 1946) is the co-founder of Infosys. He was born in Mysore and was very bright in studies from an early age. Though he cleared the entrance exam for IITs (Indian Institute of Technology), he could not join, as his schoolteacher father was unable to pay his fees. Buoyed by his father's remarks that, if you are smart enough, you do not need to study in an IIT to achieve things, he graduated in Electrical Engineering in 1967 from the National Institute of Engineering, Mysore in his hometown. Further, he obtained his master's degree from IIT Kanpur [30.1].

In spite of high paying job offers from four companies, Murthy chose to work as a Research Associate at IIM Ahmedabad, where he worked on India's first time-sharing computer system in an Institute and the third B-School in the world after Harvard and Stanford having such a system. He started a company named Softronics and after a year and a half, closed it and joined Patni Computer Systems.

In 1981, Murthy along with six software professionals started Infosys, with an initial capital of $150 (INR 10,000) given by his wife Sudha Murthy [30.1]. Facing excessive red tape from the Indian government, Infosys was barely profitable. Nevertheless, he encouraged his co-founders to hang on. Infosys fortunes changed dramatically [2] with India pursuing economic liberalization in 1991 and the rest is history. The net worth of Infosys is $2.3 Billion (2018). Murthy served as CEO of Infosys from 1981 to 2002, as chairperson from 2002 to 2011 and retired in 2011 to continue as Chairman Emeritus. He continues his service to the society as a board member of various philanthropic foundations.

STEP 31

WHEN IN DOUBT GET OPINION: SOCIAL MEDIA FOR RESEARCH/ ACADEMICS

It is tenfold more injurious to abandon the friendship of the good,
than to incur the hatred of the many -Thirukkural couplet 450

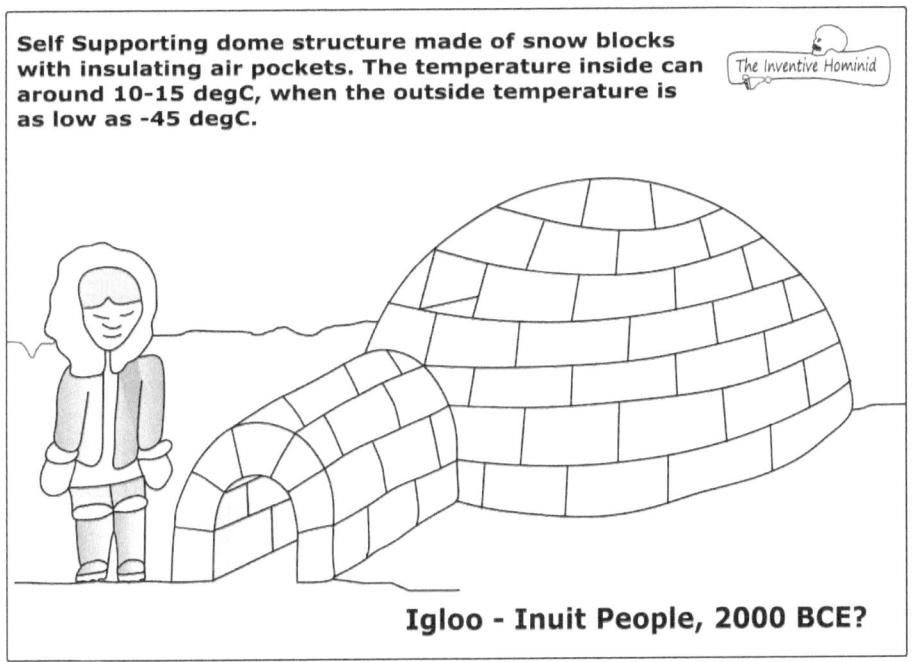

Self Supporting dome structure made of snow blocks with insulating air pockets. The temperature inside can around 10-15 degC, when the outside temperature is as low as -45 degC.

The Inventive Hominid

Igloo - Inuit People, 2000 BCE?

TIH - Igloo - Inuit (31 IR 1, 31 TR 1)

Learning Objectives

After studying the contents of this chapter, you should be able to:
- Learn about social media for professionals and researchers
- Understand various aspects for which you can get opinion on professional and academic networking sites

What will you do, when you are stuck with doubt, and want an opinion from people who have done it before? Yes, you can write to the top researchers working in this area. However, you cannot be sure about getting a reply from them. Hence, you have to cast your net wider to get the full range of opinions.

A simple way to do it is by logging onto the Q&A websites. Quora, StackExchange, LinkedIn Answers, Yahoo Answers are some of the well-known Q&A sites. Here you can ask your questions and get the answers from the accomplished people in the various tech areas.

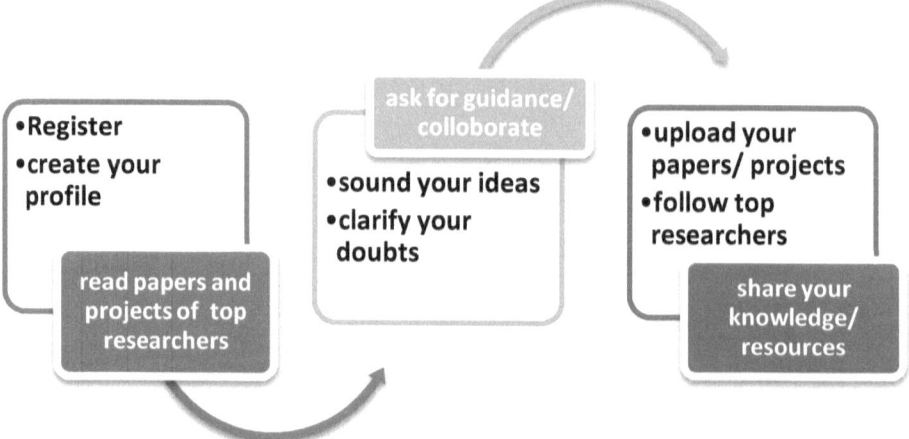

Most probably, the questions on our mind would have been answered already. By just going through the answered questions in the area of your research will get you a head start in the thinking race. In many sites, you can also ask new questions and request a well-known person in that field to answer your question. Therefore, what are you waiting for? When the whole world is itching to answer your questions, go ahead, and jump right in.

Academic Networking Sites

LINKEDIN

How about getting in touch with top industry professionals, in the area you are passionate. A site, which facilitates such a networking opportunity, is LinkedIn. Here you can get a free basic membership with which you can look at professionals from various companies and connect with them, access latest news, articles, and opinion from professionals. You can also become a member of various professional groupings and take part in discussions. Other sites, which offer professional networking, include Zing, Doostang, Makerbase (for software), efactor or Rockethub (for entrepreneurs) and many more. You can login to these sites and get an idea about the pulse of the industry, which can help put your career on the right track, a track of your passion.

RESEARCHGATE

For the ones among us who have an interest in an academic career. This site gives you an opportunity to follow the top academics in the area of your interest. You can access their publications and interact with them. Imagine having access to top professors in the world researching in the exact area of your interest. You can read their research papers and get your doubts clarified. Go ahead, discuss about your project, and seek advice when you are stuck without progress. Get to know them more by accessing their lab websites and the information given there. Other academic and research networking sites are Academia, Mendeley, etc.

QUESTIONS FOR WHICH YOU CAN GET AN EXPERT OPINION FROM PROFESSIONAL NETWORKING SITES

- What software should you use for your problem?
- What technologies should you use to build your software?
- Are there advanced alternatives to what you have planned?
- Is your circuit foolproof, and are the right components used?
- Where can you source components cheaply?
- Where can you get student discounts for software/ equipment/ components/ resources?
- Are you going in the right direction to solve your problem?
- Is your logic diagram good? How can you improve it?
- Is your problem already solved?

o   What are the potential pitfalls in your approach?
o   Is the time to finish your project sufficient?
o   Is it feasible to finish your project within this budget?
o   Where can you source funds for your project?
o   Who are the people who are experts in this domain/ area?

Inspiring Individuals

The Accidental Milk Messiah

Dr. Varghese Kurien (1921-2012) was the architect of India's White (Milk) revolution. Kurien was born in Kerala in an affluent family. He obtained his science degree from Loyola College, Madras and his Mechanical Engineering degree from Guindy College of Engineering, Madras. Eager to study abroad, he took the only available government scholarship for his MS to study Dairy Engineering in USA. After his return, he was assigned to the government dairy plant at Anand, Gujarat. Kurien hoped to just while away his mandatory service period at Anand and get out of that place.

After his service period, when he was about to leave Anand, he was approached by Tribhuvandas Patel who asked his help to setup dairy equipment for the co-operative he was trying to form. At that time, intermediaries for a Bombay based milk product company [31.1] were exploiting the milk-producing farmers. Kurien decided to stay and help for a short time, but ended up staying all his life.

With utmost dedication, he setup a modern plant and with the help of his friend Dayala [31.2], succeeded in producing condensed milk and milk powder from buffalo milk, a feat deemed impossible by western experts. Out of his untiring efforts grew Amul and 16100 milk co-operatives pushing India to number one spot in milk production. Let us thank him whenever we drink milk [31.3].

STEP 32

Effective Thought Process (ETHOP): Brain Spark – The Myths

Anyone who claims that the brain is a total mystery should be slapped upside the head with the MIT Encyclopedia of the Cognitive Sciences. All one thousand ninety-six pages of it. -Tom McCabe

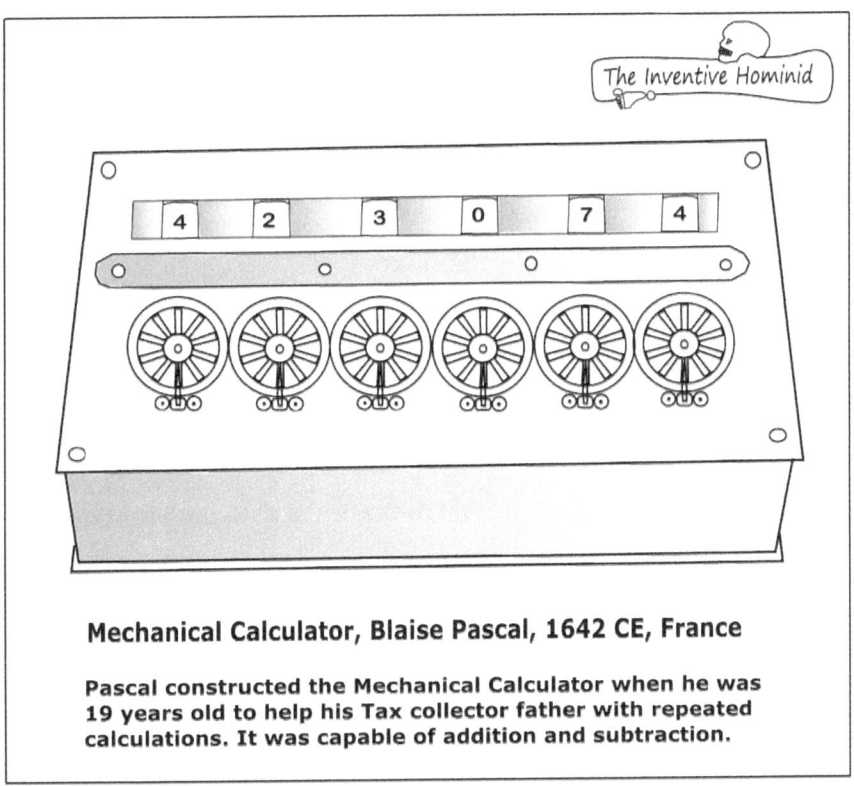

Mechanical Calculator, Blaise Pascal, 1642 CE, France

Pascal constructed the Mechanical Calculator when he was 19 years old to help his Tax collector father with repeated calculations. It was capable of addition and subtraction.

TIH - Pascal Calculator (32 IR 1, 32 TR 1, 32 TR 2)

Learning Objectives

After studying the contents of this chapter, you should be able to:
- Understand the rigour necessary to get eureka invention moments
- Learn the process of meticulous reading to enhance knowledge before looking for brain sparks

A myth ingrained in the minds of young people is that, most inventions have come through a sudden brain spark in the minds of scientists. This is both true and false. It is true that scientists get brain sparks. However, it would invariably be after loading up their grey matter with tons of information over long periods.

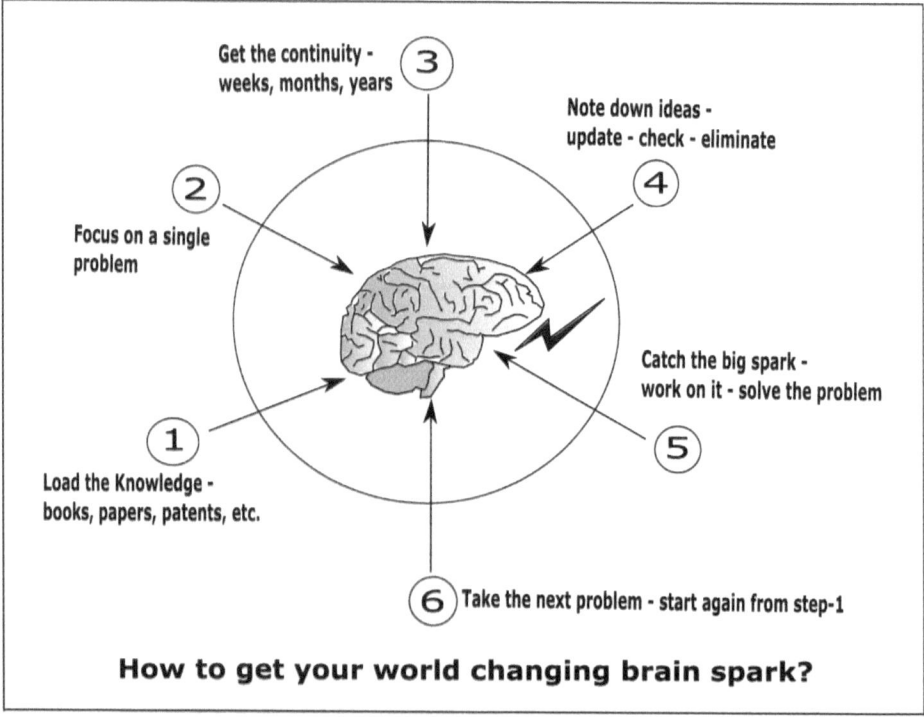

How to get your world changing brain spark?

Hence, anybody can get the "eureka" moment after they sufficiently load their brain with enough knowledge about the problem they are solving. Once up to the brim with information, our powerful brain does a good job of coming up with various combinations of solutions. Once you spend a considerable amount of your time in thinking about your problem, your brain will work on the problem even when you are asleep. Hence reports of researchers waking up from their dream with an ingenious solution, such as Double helix.

Inspiring Inventors

The Lawyer who reformed robbers

Ardeshir Godrej (1868-1936) is an Indian inventor and founder of the famous Godrej Company. Upon completing his law degree, Godrej took up a court case in Zanzibar. While he was arguing the case on his client's behalf, unable to believe his client's version fully, he refused to support his client's version in court. Thus, he gave up the law profession convinced that he cannot practice law, which requires single-minded support for the client [32.1, 32.2].

Unwilling to ask money from his father to start a business, he took a loan from Merwanji Cama and started a surgical equipment manufacturing business. When the client refused permission to stamp the products as made in India, he closed the business.

Reading in the newspaper about frequent robberies in Bombay, he decided to manufacture the safest locks. Thus came the Godrej Locks, which are till today considered the safest, and robbers of Bombay and across India, were put out of business and had to reform their ways. Godrej went on to produce a number of unique products such as burglar & fireproof safes, detector and Gordian locks [32.1].

After retirement in 1928, he took to farming in Nashik. Upon learning that the soaps were made of animal fat, which could not be used by vegetarians, he set out to produce soap out of vegetable oil. He succeeded in his endeavor that everyone thought was impossible.

Today the Godrej group has forayed into Consumer products, appliances, agri-business, real estate, etc. Godrej was one of the first companies to establish the innovative spirit of Indian companies in spite of British efforts to destroy everything Indian.

STEP 33

ETHOP - GET THE CONTINUITY

In deep meditation, the flow of concentration is continuous like the flow of oil -
Patanjali, Mahabhasya

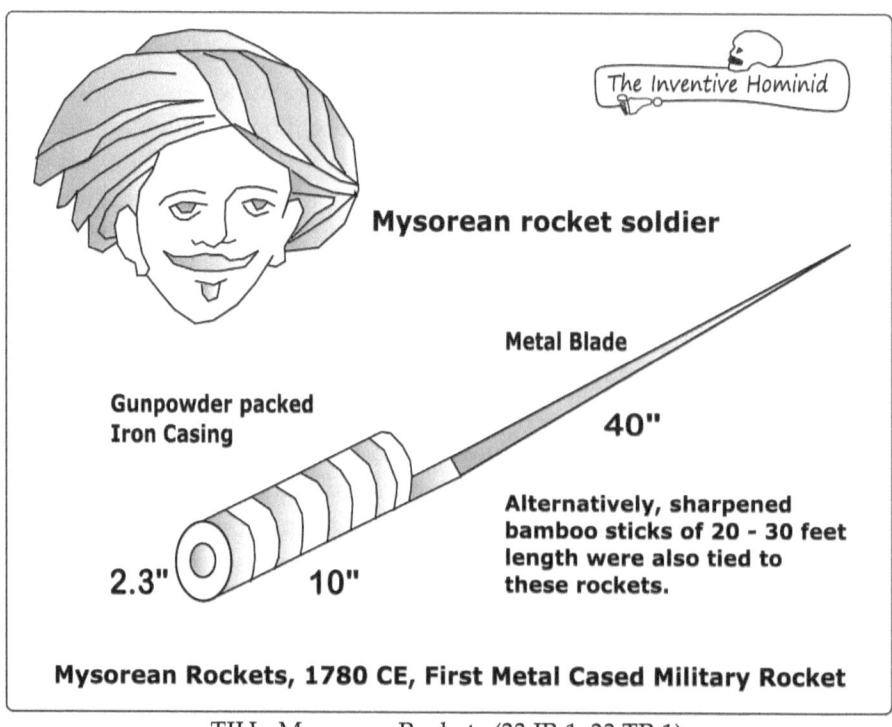

The Inventive Hominid

Mysorean rocket soldier

Metal Blade

**Gunpowder packed
Iron Casing**

40"

**Alternatively, sharpened
bamboo sticks of 20 - 30 feet
length were also tied to
these rockets.**

2.3" **10"**

Mysorean Rockets, 1780 CE, First Metal Cased Military Rocket

TIH - Mysorean Rockets (33 IR 1, 33 TR 1)

Learning Objectives

After studying the contents of this chapter, you should be able to:
- Understand the importance of continuity in thought process to achieve breakthroughs in knowledge
- Learn about the ways of great thinkers and inventors in maintaining continuity

Hours, days, months… years!

Eat, sleep, dream, walk, think about your idea…, and then watch the sparks fly in your brain.

MAINTAINING CONTINUITY IN YOUR THOUGHT PROCESS

An important ingredient to achieve breakthrough in your thought process is the critical aspect of maintaining continuity in your thinking process. Great researchers are known to be absent minded, not without a reason. They cease to live in their physical surroundings and are often deeply lost in their thought universe.

It is important that you maintain a particular line of thinking unbroken for however long it takes to satisfy us that we have exhausted all possibilities in that line of thinking. Edison is known for his long hours of experimenting stretching well into days with a few hours of shuteye now and then, right on the workbench.

Great scientists, writers, and philosophers were known to have spent their time in complete isolation alone with their thoughts. Tagore was known to emerge from his locked room after days of writing. The point, which we can understand, is the importance of tackling a line of thought to its logical conclusion. Without continuity, a line of thought will be broken many times without us being able to pick it up again. Frequent interruptions will not lead to any fruitful conclusion in the objectives of our project.

Continuity of thought process is the sole criteria for the formation of residential centres inside universities and research organizations. Of late, Tech companies have caught on to this idea by providing all facilities to their employees to make them all stay in the workplace for longer periods of time and be more productive in their research or invention process.

Hence, try to create a space for your uninterrupted thinking. This can be over the weekend or a longer time during semester breaks. A complete withdrawal from the society, including the social media will enable you to accomplish your logical thinking tasks fruitfully in much lesser time than it would with frequent interruptions.

Inspiring Inventors

The automobile inventor who graduated in Engineering at 19
33.1, 33.2

Karl Friedrich Benz (1844-1929) was a German automobile engineer who created the first practical automobile. Benz lost his father when he was two and despite poverty, his mother strived to educate him well. At age 15, he cleared the entrance exam and joined mechanical engineering at University of Karlsruhe to graduate in 1864 aged 19.

After working in several companies and many business failures, he created a two-stroke engine and patented it in 1879. Working further, he designed and patented the speed regulation system, ignition system with spark plug, carburetor, clutch & gearbox and water radiator. Soon the bank took control of his company, forcing him to resign.

He joined his friends at their bicycle repair shop in Mannheim and started a new company, Benz & Cie. After many setbacks, Benz created the Motorwagen model 2 and began selling it from 1888. The deficiencies in the automobile were rectified after his wife along with their sons undertook a long distance journey without his knowledge. Completing a 106 km trip over many steep climbs, she invented the brake lining along the way and generated immense publicity for the vehicle. Today his company still runs as the well-respected Daimler-Benz producing breathtaking automobiles with advanced engineering specifications.

STEP 34

ETHOP - Mind Focusing Techniques

The Tao that can be described is not the eternal Tao.
The name that can be named is not the eternal name.
- Lao Tzu, Tao Te Ching.

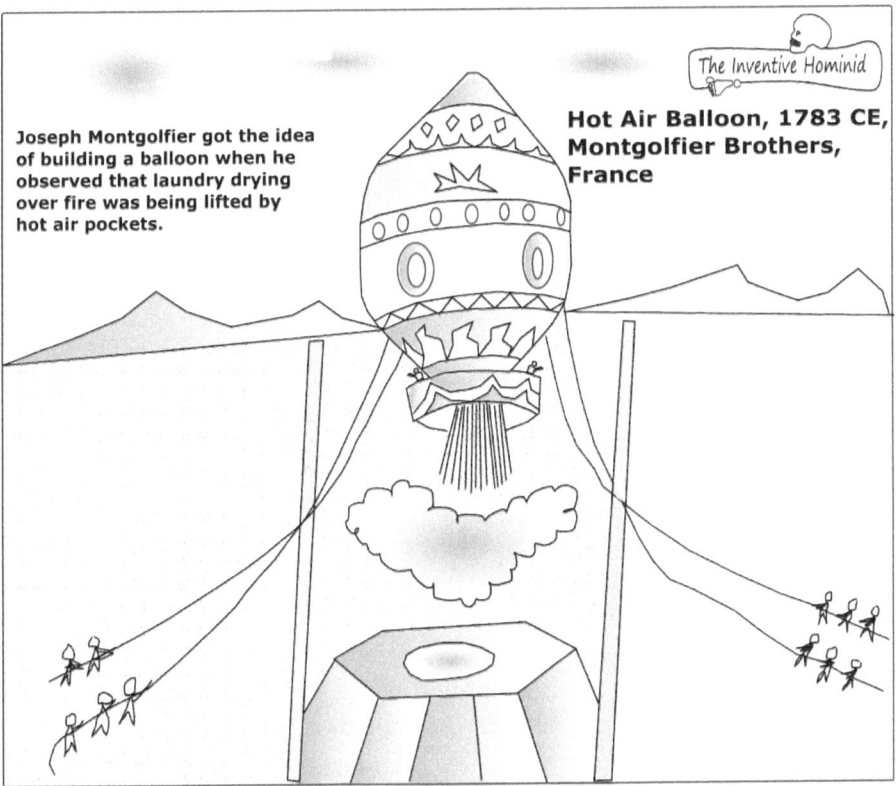

TIH - Hot air balloon (34 IR 1, 34 TR 1, 34 TR 2)

Learning Objectives

After studying the contents of this chapter, you should be able to:
- Understand the importance of maintaining a sound body for a sound mind
- Learn about various practices such as meditation, tai chi, martial arts, pranayama, music, and self-hypnosis to sharpen your body and mind

A sound body for a sound mind

What is the use of a powerful weapon, if it cannot accurately hit the intended target? Many persons waste their potential because they do not focus long enough on a problem. Adding to this problem is the plethora of social networking sites, which have cornered a major portion of the thought process of young people. So how do we train our brains to be deadly accurate weapons of effective thoughts? A number of techniques are promoted for the wellbeing of our body and mind, such as:
- Meditation
- Tai chi
- Martial arts
- Pranayama
- Music – noise blanket, mind-altering music
- Self-hypnosis

MEDITATION 34.1, 34.2

Enough has been researched and written about the benefits of meditation. Hence, it is a no brainer that meditation can help improve our concentration.

What is meditation? It is a lullaby for the brain. Why does the brain need a lullaby? Because in this modern connected world our brains are occupied with 1001 thoughts and emotions, which can exhaust us mentally. Meditation helps in calming our brain and stopping unwanted thought processes.

Imagine a computer where a number of applications are running in the background and ultimately reducing its computing power. If we can stop all the unwanted applications, which we have installed over a period, then the processor will be fast and efficient at the task, which we give it. Similarly, imagine our

neurons focusing solely on a single task and now you can understand the power of meditation.

There are a number of schools of meditation such as the Buddhist, Hindu, Chinese, Japanese form of meditation. You can learn meditation from centres, which are available all over the world. Alternatively, you can practice the basic form of meditation by yourself or by using guided tools such as audio and video clips. The basic form in all schools of meditation consists of the following steps:

- o Switch off your phone, tab, laptop, wearable devices
- o Switch on soothing meditation music
- o Sit in a comfortable position in a place of fresh air, wearing loose clothes with your legs folded if possible
- o Close your eyes and relax
- o Now breathe in slowly and deeply and concentrate on the sound of your breath
- o Slowly breathe out by concentrating on your breath

o   Repeat this for about 20 times and slowly go on to 200 times or more if possible

At the end of the session either you would have gone into deep sleep if you were sleep deprived or you will feel mentally refreshed with your brain relaxed and focused.

TAI CHI 34.3

The power of meditation

Wary of twisted yoga postures, not to worry, you can try Tai Chi. Tai-Chi! Yeah, Tai Chi or Tai-Chi Chuan is called meditation in motion. Its smooth flowing motion coupled with breathing exercise is known to reduce stress, strengthen muscles, and improve balance and respiratory capacity. Just look at some practice videos and start practicing. Of course, if you have the opportunity to learn from certified instructors and have the time and money to spare, well that should be the way to go.

MARTIAL ARTS 34.4

A variety of martial arts originating from various countries are practiced the world over. Although it was considered as an exercise in offence or defense, of late the public have realized its benefits that include enhanced physical fitness, discipline, respect, focus, and concentration. Hence, depending upon individual needs and access, one can choose from a variety of martial arts to derive its benefits.

PRANAYAMA 34.5, 34.6

Pranayama (breath control or extension) is a breathing exercise practiced as per the rules of this ancient technique from India. It consists of a series of breathing techniques designed to strengthen the respiratory system. Its benefit in our context is the infusion of oxygen rich blood into the brain and other parts of the body. Although it is claimed that Pranayama cures every imaginable disease, at the least it invigorates the mind and body by infusing fresh oxygenated blood to the brain and every tissue in the body.

MUSIC – NOISE BLANKET, MIND ALTERING MUSIC 34.7

Music has been documented as having a tremendous effect on our brains. Learning to play an instrument is believed to be a complete workout for the brain as it involves all areas of our brain pertaining to logic, motor skills, and aesthetics. It is also known to release neuron-transmitters such as dopamine, which is supposed to elevate our mood and even alleviate depression.

So grab your headphones or speakers and play your favourite genre of music. Are you not satisfied with the brain boost? Grab a guitar or a violin and start playing to give a full workout to your brain. These workouts through music will make your brain fitter and sharper and help you focus on project tasks with energy and enthusiasm.

Music can also be used in the background to cut off other noises that might disturb our concentration. Play an instrumental music and stay away from the influence of unwanted noise from outside.

SELF-HYPNOSIS 34.8

Self-hypnosis is a technique, which is used by practitioners to achieve an altered state of mind through which they gain improved focus and concentration. It is also used to block the sensation of extreme pain without the use of painkillers. In our context, it can be used by those having problems with focusing their mind and to enhance the level of concentration. It is also reported that feeding positive messages through self-hypnosis is an effective anti-dote to depression.

Effective and enhanced concentration makes a big difference in terms of understanding, memory, and reduced time to cover vast material in a subject.

However, it is advised that self-hypnosis should be attempted / learned only from certified professionals.

Create a routine and stick to it for a week and you will do it in your sleep 34.9

Great things can be accomplished when chasing excellence becomes a habit. Great achievers were able to achieve so much due to their unwavering dedication to their work routine.

Plan a routine for your daily activities based on your personality type (for example night owls tend to do better at night, whereas early risers achieve much in early morning before the world awakes). Make sure you give time to meditation and physical activity to recharge your mind and body. Plan your routine/ schedule to maximize your productivity and stick to it religiously for at least a week. After that, it will become ingrained in your body clock and you will have little difficulty in following it.

Inspiring Inventors

The habits that inventors followed to avoid distraction in their thought process 34.10

Edison was not a fan of regular sleep routine every night; instead, he just took a series of short naps in his research laboratory and dived into his inventing work whenever he woke up. Leonardo Da Vinci and Nicola Tesla are other genius inventors known to have relied on short polyphasic sleep cycles.

Many inventors including Steve Jobs were known to wear the same type of clothes every day. Roll neck Jumper and jeans in the case of Steve Jobs, which they attributed to not wanting to disturb their creative process by using their brains to decide on their dress every day.

The highest number of patent holder in the world today Dr. Nakatsu of Japan has a habit on concentrating by sitting in his toilet surrounded by the sound of flowing water with walls of Gold tiles. He insists that gold blocks all the unnecessary radio waves that might disturb his brain waves.

STEP 35

ETHOP - Creative or Lateral or Out-of-the-Box Thinking

Broken symmetry is imperfection, but rich in creativity. The universe is created out of broken symmetry -Amit Ray, Meditation: Insights and Inspiration

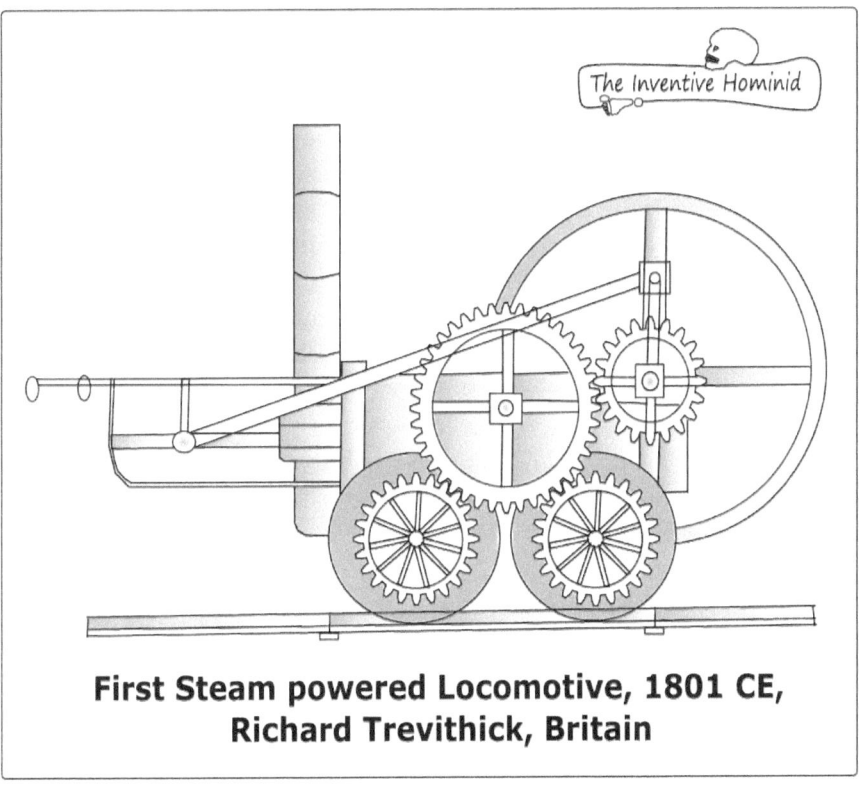

First Steam powered Locomotive, 1801 CE, Richard Trevithick, Britain

TIH - Locomotive (35 IR 1, 35 TR 1)

Learning Objectives

After studying the contents of this chapter, you should be able to:
- Learn about creative or lateral or out-of-the-box thinking
- Understand the process/ technique to power your creative thinking

Creative or Lateral or Out-of-the-Box Thinking is a highly valued asset in today's world. Who would not appreciate an ingeniously simple solution to complex problems?

What is creative thinking? 35.1, 35.2

"As a man thinketh in his heart so is he," A man is literally what he thinks, his *character being the complete sum of all his thoughts.*- James Allen

Simply put, creative thinking is about thinking about things and solutions in ways and directions that have not been attempted before. It is thinking without any inhibitions without worrying about failure of that solution.

Most people do not attempt to think creatively mainly due to their fear of being embarrassed or shamed if the suggested solution results in a failure.

This fear of failure or being shamed in front of others is a remnant of our experience in school where a wrong answer to the teacher's question was greeted with boos and laughter from our classmates.

However, the essential aspect of any endeavor to find a solution to a critical problem is by examining/ experimenting with all possible ways to a solution. Hence, it is time to understand that failure per se is not the end, but rather an elimination of an improbable step to a solution.

Here it is very relevant to appreciate Edison's statement, when somebody remarked that he has failed a thousand times to invent the right filament for the light bulb, he famously remarked that he has not failed, but has found a thousand materials which do not work well as a light filament.

Creative thinking needs a good knowledge base to start your process. Only when you have an information base about the whole gamut of materials, processes, culture, economics, etc. you can come up with a new way of combining things to provide a solution.

Your knowledge base has to be very wide about everything under the sun to be able to make the connection between seemingly unrelated things.

In many cases the adage 'necessity is the mother of invention' is what spurs innovation and creativity. This is why societies living under immense resource scarcity tend to come up with creative ways for its sustainable utilization. Modern creative thinking is all about coming up with a creative design of an existing product such as the iPod and iPhone or the creative use of newer materials such as the rapidly evolving solar cells.

Hence, to be a creative thinker, all you need to do is:

- ❖ Read widely about every topic under the sun, if you had done this from your childhood you would have covered lots of ground, but it is never late to start.
- ❖ Cast aside your inhibition about what might happen if you fail in your attempts, remember there is no such thing as a failure, as each attempt adds knowledge
- ❖ Get your basics right about various materials; ultimately, everything in this world boils down to material science

❖ If you are working in computing, then you need not worry about spoiling physical things, as all you are going to do is program and test it. Be creative and do not limit yourself to current computing paradigm.

❖ Use both hemispheres of your brain – strum a guitar, paint, practice tasks with both hands, learn a new skill/ language every year.

Inspiring Inventors

Brothers in arms – bicycle to aircraft – self taught and creative 35.3

The Wright brothers, Orville (1871-1948) and Wilbur (1867 – 1812), were American inventors who built the first controlled, sustained flight of a powered aircraft heavier than air. The Wright brothers first tried their hand at running a printing press, after which they switched to repair and sales of the Rover Safety Bicycle, which was a craze in America at that time. Later, they started manufacturing bicycles of their own brand. Reading accounts of experiments in flying by various inventors, they correctly concluded that the failures of the current efforts at flying is due to a lack of control, much like the bicycle before a beginner learns to control by practicing.

Hence, they decided to focus on the technical aspect of controlling a gliding craft. They wrote to the Smithsonian institution requesting books and articles about flying technology. They then immersed into the published literature right from Leonardo Da Vinci to Langlely and Lilenthal. After numerous setbacks and years of experimenting with aircraft models and a wind tunnel they built at home, they designed the controls of the aircraft to control it in the three axes. After many crashes and injuries, they finally managed to keep the craft airborne with control for 26 seconds covering a distance of 622.5 feet in 1902 and a controlled powered flight in 1903.

Then followed a period of struggle when nobody believed and the press called them liars. Finally, in 1908 after the successful flying demo in France that lasted for 1 min 45 seconds with amazing control to fly a circle. They became celebrities overnight, with presidents inviting them and decorated them with awards.

STEP 36

ETHOP - Note down the Sparks

One is, sit alone. In your arrangements, for your residence see that you have a chamber to yourself, though you sell your coat and wear a blanket. The other is, keep a journal. Pay so much honor to the visits of Truth to your mind as to record those thoughts that have shone therein. -Ralph Waldo Emerson, "The Head,"

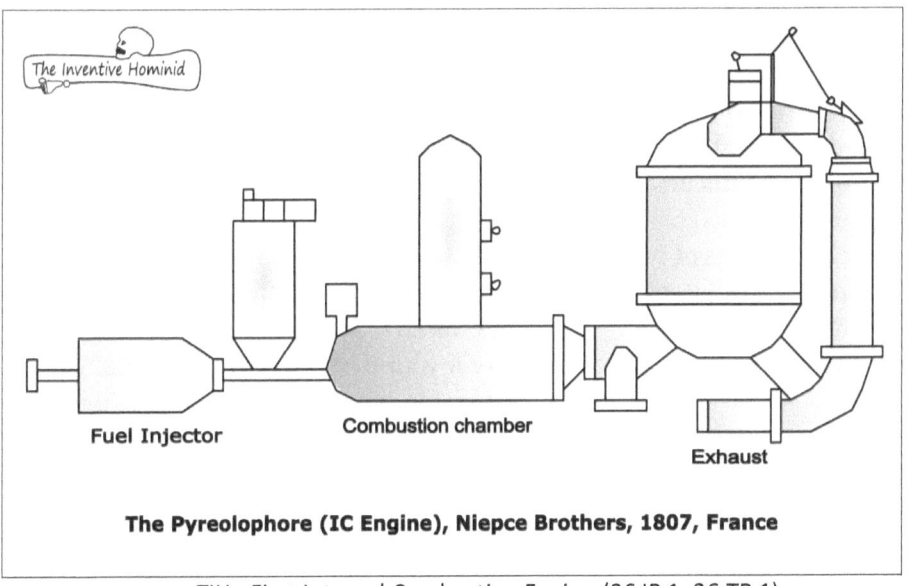

The Pyreolophore (IC Engine), Niepce Brothers, 1807, France

TIH - First Internal Combustion Engine (36 IR 1, 36 TR 1)

Learning Objectives

After studying the contents of this chapter, you should be able to:
- Learn about the importance of maintaining a notebook to note down your thoughts
- Know about the notebooks of great inventors available online

Always Carry a Notepad

An important aspect of keeping track of our thinking process is to take note of the ideas that our brain comes up with every now and then. It is essential that you should always carry a notepad with you. Use of electronic gadgets including our smart phones for taking notes will not be efficient. By the time you use their functions to note down what is in your mind, the idea would have disappeared altogether.

A great note-taking tool will be a plain pocket size notebook and an eraser tipped pencil. Notes can be in the form of words, signs, diagrams, etc. A pencil to note down the bizarre ideas that sprout from our neurons with diagrams and flowcharts is a great idea. It makes for interesting reading days or months from the time, we noted it down.

A properly maintained notebook containing your ideas and inventions are considered **legal documents**, which have helped inventors to prove their claims [36.1],

Look at the notes of the legendary Leonardo Da Vinci (The Codex Arundel) on the internet and you will be astonished at the sheer details of his drawings of things ranging from flying machines to the anatomy of the human body (www.bl.uk/manuscripts). Do not worry if your drawing skills are not as good as Da Vinci, it would be the case for the most in our world. Just draw without any inhibitions; it should just be good enough for understanding by you after days or months. Combine words, diagrams, sketches, lines, mind maps, and anything, which comes to your mind.

Notes help you capture intense activity in your brain, which suddenly explodes into action at unexpected moments. The only way to capture those sparks is by noting them down and then you can leisurely digest the information you have captured at your convenience.

Inspiring Inventors

"and yet it moves", you can now read the notebooks of the 'father of modern Science' 36.2

Galileo Galilei (1564-1642) was an Italian Astronomer, Physicist, and Engineer. He is regarded as the father of modern science, observational astronomy, modern physics, and the scientific method. Galileo started studying medicine as per the wish of his father, as physicians earned more than mathematicians & scientists even in those days. However, hearing a lecture on mathematics, Galileo changed his mind and requested his father to let him study mathematics and natural sciences.

Based on his astronomical observations with a telescope, he challenged the then existing belief that the Earth was the centre of the universe, and all heavenly bodies including the Sun revolved around the Earth. He was accused of heresy by the Roman Catholic Church and was asked to refute his theory. Unable to hide the truth, he recanted and ended with the statement "and, yet it moves'. Angered by his stubbornness, he was sentenced to house arrest and his books were banned.

Restricted to his house, he wrote two of his greatest books on 'Two new Sciences' now called as Kinematics and Strength of Materials. His notebooks containing his theories and sketches can be viewed online at:
https://www.museogalileo.it/en and
https://brunelleschi.imss.fi.it/bibliotecagalileo/sez_3.html

STEP 37

Power your Project with Open-source Tools

I want to encourage free software to spread, replacing proprietary software that forbids cooperation, and thus make our society better.
- Richard Stallman, Copyleft: Pragmatic Idealism

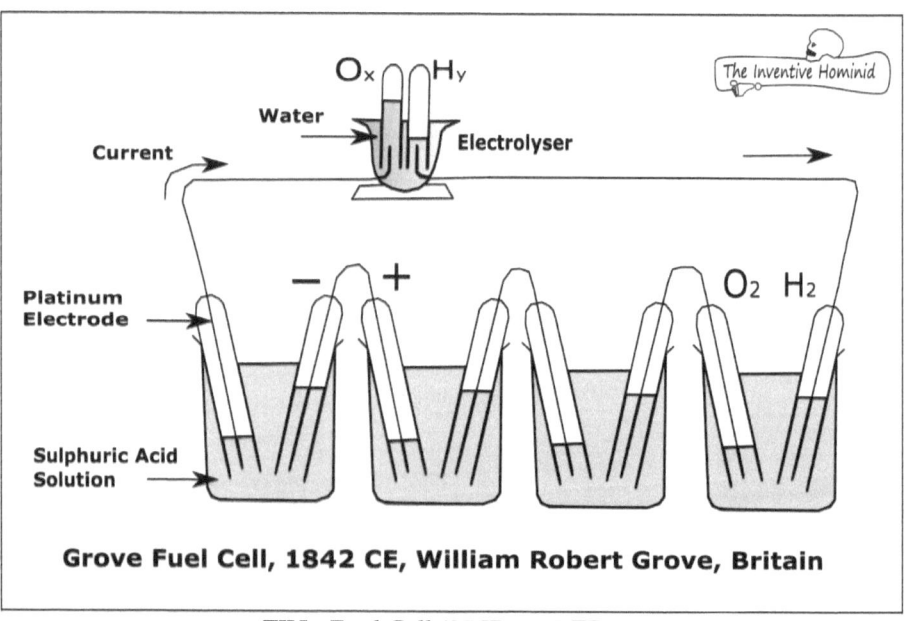

TIH - Fuel Cell (37 IR 1, 37 TR 1)

Learning Objectives

After studying the contents of this chapter, you should be able to:
- Understand the importance of software tools to enhance your engineering project outcomes
- Know about open-source free software available for every aspect of an engineering project

Open Source Engineering Tools – Turbo Charge Your Project   37.1, 37.2, 37.3

- ❖ Coding Tools
- ❖ Simulation Tools – Virtual Labs
- ❖ Design Tools – Arduino
- ❖ Reading and Writing Tools
- ❖ Grammar Checker

You can turbo-charge your project with open source software tools and save a lot of time and be efficient and accurate in your simulations and calculations.

Open source refers to the software whose source code is available for free download, modification and usage. Most licensed software offer free student versions, which you can use if you are not able to access the licensed version.

These tools are available for every domain of engineering. All you have to do is browse around the web by searching with key words and there you have it the latest tools ready for download and use along with tutorials. The names of websites, tools, products, and companies are registered/ copyrighted names that belong to the respective companies, and shall not be used without permission from the respective owners. Further information can be availed from their websites.

PROGRAMMING/ CODING TOOLS

Programming is an attribute that is necessary in all engineering domains, which involves repetitive or excessive computation. Hence, it is very much desirable that engineers upgrade their skills by learning to program. For most of your problems, you might just use a software package that lets you do your project without having to program or write a code. However, in certain situations you might not get the functionality you require and hence would have to write the code yourself.

Learning to write a code or program is a nightmare for most engineering students, including those studying computer sciences. Considering the advance of technology that has resulted in a ballooning curriculum, there is hardly time for the teacher to go slow or worry about the stragglers in a programming course. However, since the advent of the internet and the plethora of online courses and tools, all you have to do is manage your time efficiently and focus on learning online.

Some popular coding websites are listed below. It is advised that searching for the top coding websites on the internet when you are ready to learn is a better way to get the latest updated list which will exclude sites that might become defunct over time.

- Codeacademy
- Code school
- Code house
- Learn street
- Udacity
- Khan Academy
- Scratch 2.0
- SQLZoo
- Code Racer
- General Academy
- PHP Academy
- Coding Bat
- Hackety Hack

ENGINEERING DESIGN TOOLS

Design tools are those that help engineers to visualize/ simulate and prepare the final schematics of the product they are designing. There is a range of software and apps available that enable engineers to transform their ideas into products. Your institute would have the licensed version of most of these softwares in your design lab. If it is not available, you can always use the trial version or request for student version that most companies are happy to provide.

Some of the tools include:

- Engineering cookbook – mechanical engineering reference guide – app – Google play
- AUTODESK product design suite – for 3D product design.
- Scilab – numerical computation for engineering and scientific applications

- MATLAB – for numerical computation, visualization, and programming
- HyperWorks – modeling, visualization, and design.
- Fusion 360 – combined mechanical design and simulation.
- Revit – Building Information Modeling
- STAADpro – Structural analysis and design
- NX for Design – 3D product design - software – Siemens
- GeckoCIRCUITS – circuit simulator – software
- Autodesk Digital Prototyping – software - 3D product design
- XCircuit – circuit schematic diagrams – software
- EAGLE PCB Design – software
- CATIA V5 – 3D design, simulation and analysis.
- QCAD – 2D CAD software
- Solidworks – software – 3D modeling
- INOVATE – software – collaborative 3D modeling and visualization
- NASTRAN – FEA simulation and modeling.
- CircuitLab – software – circuit simulation and design
- TinyCAD – software – circuit and PCB layout drawing
- FreePCB – software – PCB design
- Femap – software – finite element analysis modeling
- MapleSim – physical modeling and simulation
- ANSYS – modeling flow, turbulence, heat transfer, turbo machinery.
- Geomagic Design – 3D modeling
- MechDesigner – tool for design and analysis of machinery
- CATIA – tool for product design and innovation
- Actran – Acoustic simulation tool
- SCIA Engineer – Structural analysis, design and BIM modeling
- SketchUp – 3D drawing for architecture and construction
- EPA Net – water distribution network analysis and design

OPEN SOURCE PLATFORMS FOR HARDWARE DESIGN

The development of hardware through open source platforms for a variety of applications ranging from a radio to 3D printers to automobiles has gathered momentum over the years. You can use these projects to improvise, modify, and develop your own product. There is no need to reinvent the wheel and waste precious time. You can find a comprehensive list in Wikipedia – List of open source hardware projects. A few of the Open source Hardware projects are given below:

o Farming Tools – FarmHack, FarmBOT
o Communication System – Pirate Box
o Video – Milkymist One, Neuros OSD
o Wireless networking – NetFPGA
o Telephony – OpenBTS, Project Ara, Openmoko
o Robotics – ArduCopter, ICub, OpenROV, IOIO, OpenRAVE, Spykee, Thymio, Tinkerforge
o Homes – Wikihouse
o Furniture – Opendesk
o Vehicles – OSVehicle, Rally Fighter, Oscar, Wikispeed, OpenXC, OpenEVSE, SECU-3.
o 3D Printers and scanners – RepRap, LulzBot
o Entire Village – Open Source Ecology
o Building – Openstructure
o Electronics – Nitrokey, The Bus Pirate, NodeMCU, Arduino
o Computer systems – CUBIT, Arduino, Netduino, Chumby
o Cameras – AXIOM, Elphel
o Drone – APM Copter
o Modular Smartphones – Project ARA

ONLINE RESEARCH TOOLS
WRITING AND REFERENCING TOOLS 37.4

o Endnote – referencing
o Writecheck – plagiarism and grammar checker
o Thesis statement creator
o Purdue owl – writing, grammar, plagiarism check, styles
o Gliffy – data into flowcharts and diagrams
o Citeulike – referencing

RESEARCH SEARCH ENGINES/ PLATFORMS 37.5, 37.6

o Google scholar
o Google books
o Directory of open access journals
o Citeseerx
o Science direct
o Worldwide science
o bioone
o Science.gov

- Research gate
- Base
- Pubmed central

GRAMMAR CHECKING

Some of the popular Grammar Checkers include:
- Whitesmoke
- Scribens
- Grammarly
- Reverso
- Ginger
- Prowritingaid
- Sentence Checker
- Spellcheckplus

Unable to find free versions for your need? Why do not you try the Linux versions from the GNU Project?

GNU Project

It is a free-software, mass collaboration project, started in 1983 by Richard Stallman at MIT. It aims to develop free-software by collaboration and give it to users who can run it, share, and modify it with freedom.

Linux

Linux is a family of free and open source software systems based on the Linux Kernel. Linus Torvalds released the Linux kernel, an OS Kernel in 1991. You can use the Linux OS freely instead of paid Windows OS. The Android OS for smartphones is also based on the Linux kernel.

FREE SOFTWARE ALTERNATIVES TO PAID VERSIONS

You can find free versions for almost every need in the directory of the Free Software Foundation. The directories are located at:

For Linux Version: https://directory.fsf.org/wiki/GNU

For windows version: https://directory.fsf.org/wiki/Collection:Windows

Open Source Software Directory for all platforms:

https://alternativeto.net/software/open-source-software-directory/

Note down interesting tools for learning and use in your project:

_____

_____

Inspiring Inventors

The mind behind Linux, which has taken over computing

Linus Benedict Torvalds (born 1969) is a Finnish-American software engineer and the principal developer of the Linux kernel. He attended the University of Helsinki and after the first year, joined the Finnish Army for a year to fulfill the compulsory military service of Finland. Later, he graduated with a master's degree in Computer Science with a Thesis titled Linux: A Portable operating System [37.7, 37.8].

Interested in computers and programming from childhood, he was modifying the OS of the computers available to him. While at the University, he publicly released the first prototype of Linux in 1991 and the version 1.0 in 1994.

The interest of developers in Linux quickly spread throughout the world and in 1999, Red Hat and VA Linux, presented Torvalds with stock options in recognition for creating Linux. When both companies went public, Torvalds' share value raised up to US$20 million. Since then Linux has conquered all but the desktops and Linus has been decorated with numerous Awards [37.9].

STEP 38

Collaborate with others Worldwide and Get Online Research Mentors

Great things are not done by impulse, but by a series of small things brought together
-Vincent Van Gogh

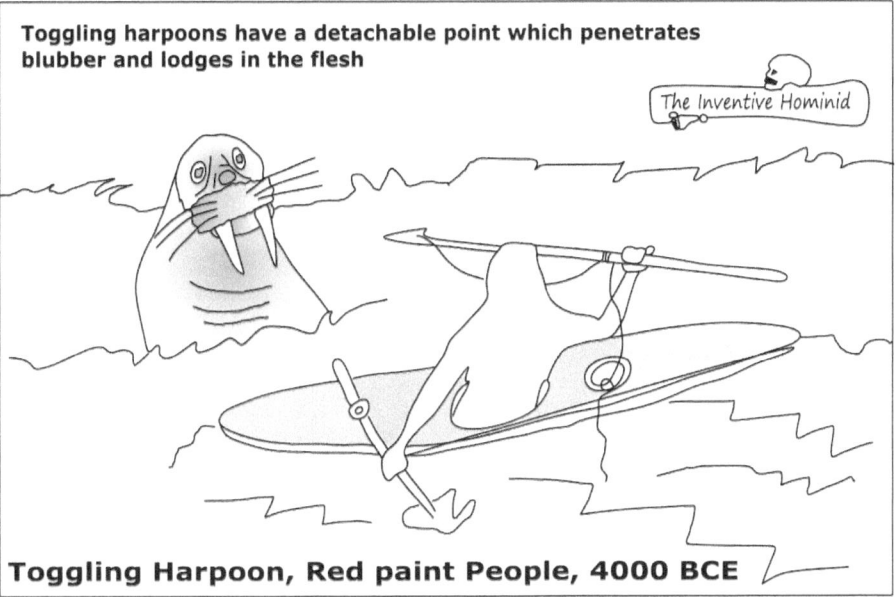

Toggling Harpoon, Red paint People, 4000 BCE

TIH - Steam Hammer (38 IR 1,38 TR 1)

Learning Objectives

After studying the contents of this chapter, you should be able to:
- Learn about academic research websites and the networking aspect of them
- Learn about ways to collaborate and get mentorship through various websites

Are you worried that your college does not have research-oriented faculty? Do you want to sound your ideas with an expert? On the other hand, do you need a mentor in an emerging area of research? Worry not, for you can simply go online and get support from around the globe.

ACADEMIC NETWORKING SITES

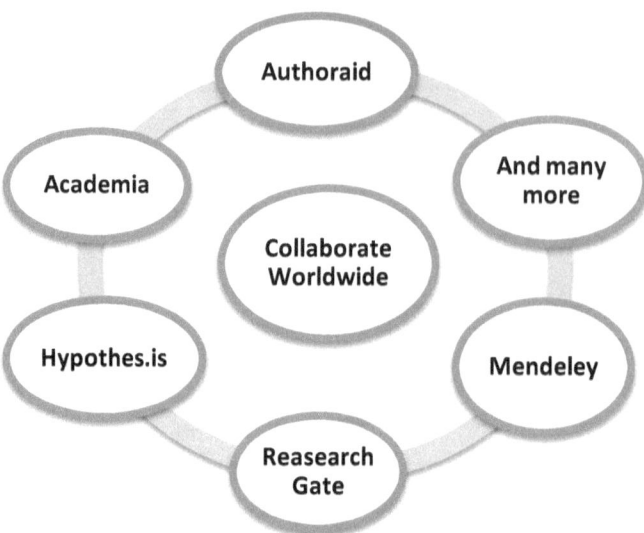

Create your account in any of these academic networking sites [38.1] like
- ResearchGate
- Academia
- Mendeley
- Authorea
- Diigo
- Figshare
- Frontiersin
- Hypothes.is

In addition, just explore the work of top researchers in the area you are working on, ask questions about their work and get your doubts clarified.

On these sites, you can explore the areas in which the leading researchers are working on, read the papers published by them over the years, discuss about their published work after reading their papers, get an idea about popular journals in various fields, and get a concrete idea about which way the research trends are moving.

Then there are sites like **Authoraid,** which allows you to connect with a mentor from among researchers across the world. Now what are you waiting for, just dive into the sea of intellect and swim to your destination.

A comprehensive directory of academic networking sites can be found at [38.1]: https://www.timeshighereducation.com/a-z-social-media

Inspiring Inventors

The man who overcame threats and ridicule to innovate for poor women

Arunachalam Muruganantham (born 1962) is an inventor from India known for inventing the low cost sanitary pad-making machine. He overcame personal abuse to break the taboo surrounding menstruation in India. His machines are installed all over India providing low-cost pads at 30% of price of pads available in the market [38.2].Muruganantham born to handloom weavers grew up in poverty after his father died in an accident. He worked as a farm laborer while studying in school and dropped out at the age of 14 to work full time. He worked in various jobs to support his family.

After marriage, noticing that his wife was not using sanitary pads, as they were not able to afford them, he decided to find a low cost way to make them. He started experimenting with various materials, but could not convince anybody to try them. He started testing himself with a bladder filled with animal blood.

After two years of experimenting, he found out that commercial pads were made of cellulose fibers derived from pine wood pulp. He sourced pine wood pulp from Mumbai, created a machine to grind, de-fibrate, press, and sterilize the pads, before packing them. The cost of his machine was INR 65,000 (USD 915) as against INR 3.5 Crores (USD 500,000) for imported ones [38.3].

Refusing to sell his machine to companies, he is manufacturing them through his company named Jayaashree Industries [38.4] to distribute among rural women self-help groups with government support. He has plans to export these machines to countries around the word. He was awarded Padma Shri by Indian Government in 2016.

STEP 39

Pitch Your Idea & Flex your Neurons through Competitions

The important thing is not to stop questioning. -Albert Einstein

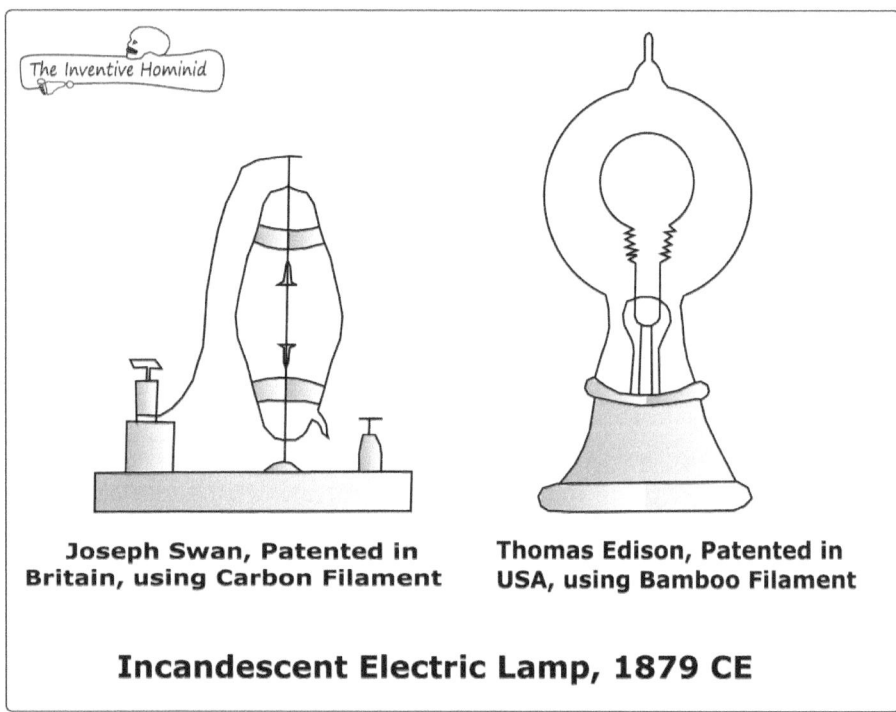

The Inventive Hominid

| Joseph Swan, Patented in Britain, using Carbon Filament | Thomas Edison, Patented in USA, using Bamboo Filament |

Incandescent Electric Lamp, 1879 CE

TIH - Electric Lamp (39 IR 1, 39 IR 2, 39 TR 1)

Learning Objectives

After studying the contents of this chapter, you should be able to:
- Learn about various engineering competitions around the world
- Understand the importance of competing to enhance your engineering project

A great way to strengthen the neurons in your brain is by taking part in design competitions. Prepare a list of 3-5 national/ international competitions per semester/ year and Compete among the best in the world.

ENGINEERING COMPETITIONS

- Building a Lunar Rover
- Designing a Nanobot for Vascular Surgery
- Real-time Translator for Smart Phones
- Remote Operated Home Appliances
- Smart and Sustainable Buildings
- Zero Energy Transport Systems, and much more.

Target annual competitions and start preparing right away without waiting for the announcement

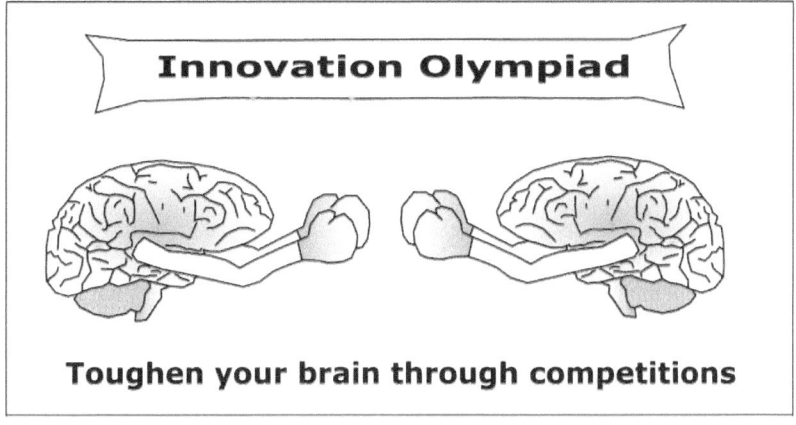

POPULAR DESIGN COMPETITIONS 39.1

- Reduced Gravity Flight – NASA
- Aircraft/ Engine/ Space Design - American Institute of Aeronautics and Astronautics
- Go green in the city – Schneider Electric
- Shark Tank Wellness – Global Wellness Summit
- Carbon Footprint Challenge – UNITECH
- Taiwan International Student Design – Taiwan Ministry of Education
- Chemical Powered Car – AICHE – American Institute of Chemical Engineers
-

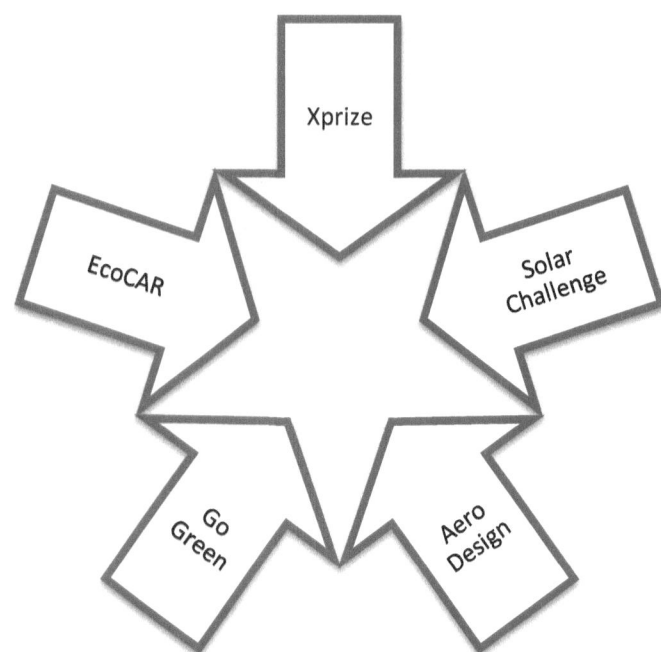

- o Human powered helicopter – American Helicopter Society
- o ¼ Scale Tractor Design – American Society for Agricultural and Biological Engineers
- o Vertical Flight - AHS International Annual Student Design Competition
- o Ecocar2 – US Department Of Energy and General Motors
- o Solar Decathlon – US Department Of Energy
- o International Student Tall Building Design Competition - The Council on Tall Buildings and Urban Habitat (CTBUH)
- o ASME Prototype Design – American Society of Mechanical Engineers
- o Concrete Canoe – American Society of Civil Engineers
- o Aero Design – Radio Controlled Aircraft – Society of Automotive Engineers
- o Formula SAE – Society of Automotive Engineers
- o Design for Direct Digital Manufacturing – Society of Manufacturing Engineers
- o World Solar Challenge – South Australian Motor Sports Board

Most of these societies maintain a web based database of past competition and winning designs, go through them, soak up the finer aspects of winning designs, and try to match those standards through focused hard work. Then you are sure to win not one, but a number of competitions.

Inspiring Inventors

The merchant of death who instituted Nobel Prize
39.2, 39.3

Alfred Nobel (1833-1896) was a Swedish inventor known for inventing the explosive called dynamite. His father was an Inventor with the veneer lathe among his inventions. Private tutors taught Nobel and he excelled in chemistry and languages. He attended school only for eighteen months. While still in his teens, he started working with chemist Nikolai Zinin. Then at the age of 17, he went to Paris, where he met the inventor of nitroglycerin Ascanio Sobrero. Sobrero opposed the use of nitroglycerine, as it was dangerously unstable. However, Nobel resolved to find a way for its safe use as an explosive.

At the age of 18, he went to US to study and worked briefly under inventor John Ericsson, who designed the ironclad USS Monitor. Returning to Sweden in 1859, Nobel devoted himself to study and improvement of explosives. He invented a detonator in 1863 and an accident at their family armament factory in Stockholm killed five people, including his younger brother Emil Nobel. Unfazed, Nobel continued his work of building more factories and invented a blasting cap in 1865, dynamite in 1867, gelignite in 1887 and ballistitie in 1887. Nobel in his lifetime was issued 355 patents and established more than 90 weapons factory, despite his declared belief in pacifism.

When his brother Ludwig died in 1888, newspapers mistook it for his death and published obituaries. One French newspaper published an obituary titled 'The merchant of death is dead'. Depressed at being called the merchant of death, Nobel decided to change the way people perceived him by leaving 95% of his wealth for the Nobel Prize. Until date, Nobel Prize is awarded each year for significant contributions in Physics, Chemistry, Physiology, Literature, and Peace. The Indians who have received the Nobel Prize include Rabindranath Tagore for Literature, CV Raman for Physics, Mother Theresa for Peace, Kailash Satyarthi for Peace and Amartya Sen for Economics. The Swedish Central Bank in memory of Alfred Nobel instituted economics prize

STEP 40

Project Funding – How to raise money for your project?

There are three faithful friends, an old wife, an old dog, and ready money - Benjamin Franklin, Poor Richard's Almanac (1734).

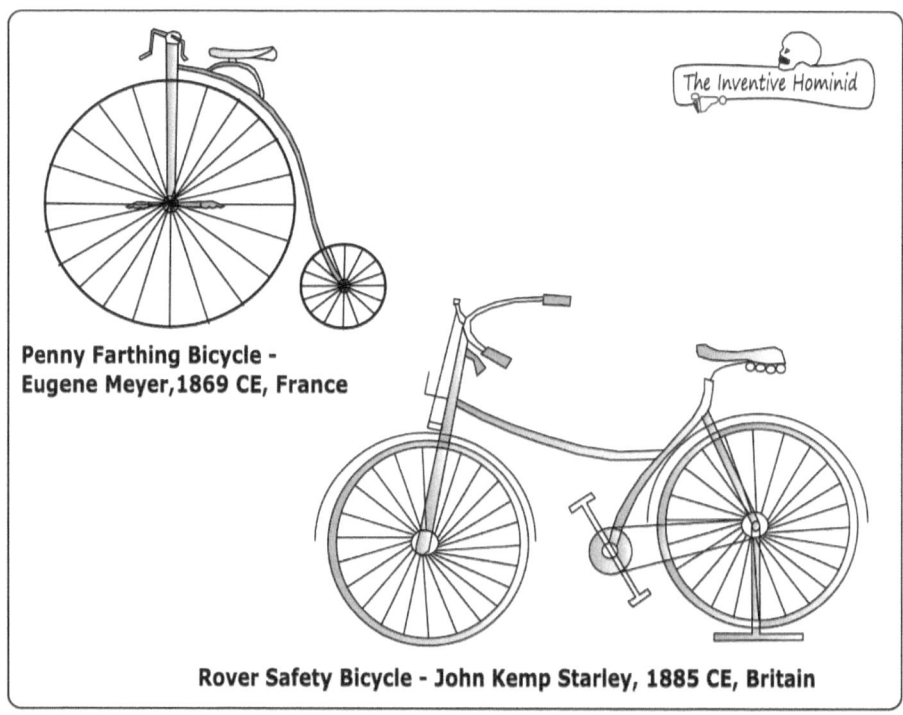

Penny Farthing Bicycle -
Eugene Meyer,1869 CE, France

Rover Safety Bicycle - John Kemp Starley, 1885 CE, Britain

TIH - Bicycle (40 IR 1, 40 IR 2, 40 TR 1)

Learning Objectives

After studying the contents of this chapter, you should be able to:
- Learn about various avenues available to get funding for your project
- Understand how to write a research project proposal to get funding

Now that you have chosen the problem and started to sketch your solution, it is time to think about the expenses. When you expect funds from somebody, you need to provide details of your project and convince them. We now explore the various options available to raise the funds needed for the project and the details required to present your project idea in a convincing manner.

Self Funding – Bootstrapping 40.1, 40.2

List out the things you need to complete the project – components, software, computing resources, testing facility, etc.

Check whether these are provided in your institution. Most of the institutes provide you with computing and testing facilities. You might have to buy the necessary components and consumables to build, test, and execute your project.

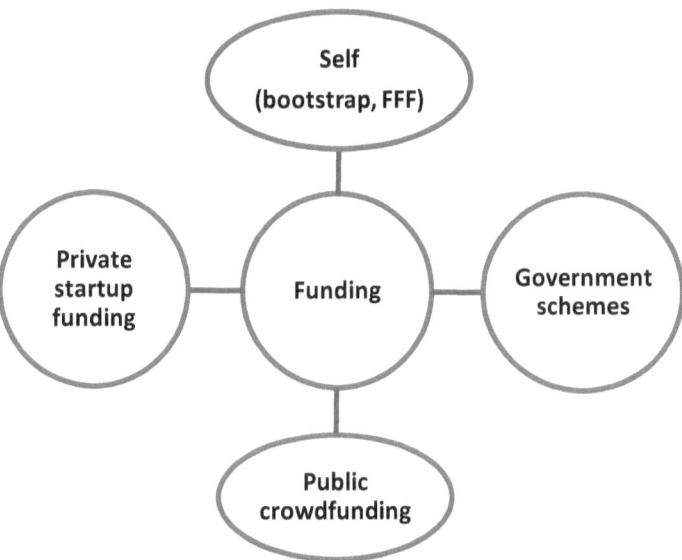

Check out the offers on the internet for various components that you can now buy from around the world at competitive prices.

One way to fund is by bootstrapping i.e. self-starting with whatever funds you can manage on your own.

Then you have the **FFF (friends, family, and fools)**[40.3] - you can convince people who know you, like friends and family or fools (people who have the money and do not give much importance to it) who can fund you based on the trust they have in you, if not on your idea.

Public Funding (Government Agencies, University, etc.)

You can apply for funding from your University, Government Agencies, Professional Associations, Private Trusts, etc. These sources mainly fund research projects that advance our knowledge.

Do a thorough internet search for student funding in your country and make a list of agencies offering research funds and the deadlines for applications.

Normally you need to have a detailed research proposal, which outlines your research objectives and outcomes clearly, based on which your proposal can be evaluated for funding.

PROJECT RESEARCH PROPOSAL 40.2

The research proposal is a document that details the idea that you want to turn into a research project. Reviewers decide whether your idea is worth providing funds by evaluating this document.

The elements of a research proposal are:
- o   Title of the research project
- o   Abstract of your proposed research
- o   Background of the research proposal – it contains a detailed literature review to show where the problem/ idea stands currently. This is where you have to show a ton of references from published papers and reports
- o   The introduction or the origin of your project/ product idea or your research hypothesis (if it is a research-oriented project)
  Hypothesis refers to the central premise of your research that you will prove or disprove at the end of your project.
- o   Relevance of your research idea to the goals/ objectives of the funding agency/ nation/ world
- o   Clear-cut objectives of your research proposal
- o   Well-defined outcomes expected

Establish the need for research through rigorous literature review (refer chapter 28 Literature Review)

A meticulous literature review is half the job done. Literature review gives you the information about research that has been done till now on the problem you have chosen; what are the results gained till now; what are the problems (research gaps) which still persists and which direction is the overall thinking headed.

With the above information, you will be able to unleash your thinking on what other directions you can move the research on. Choose a research gap/ problem as your research topic.

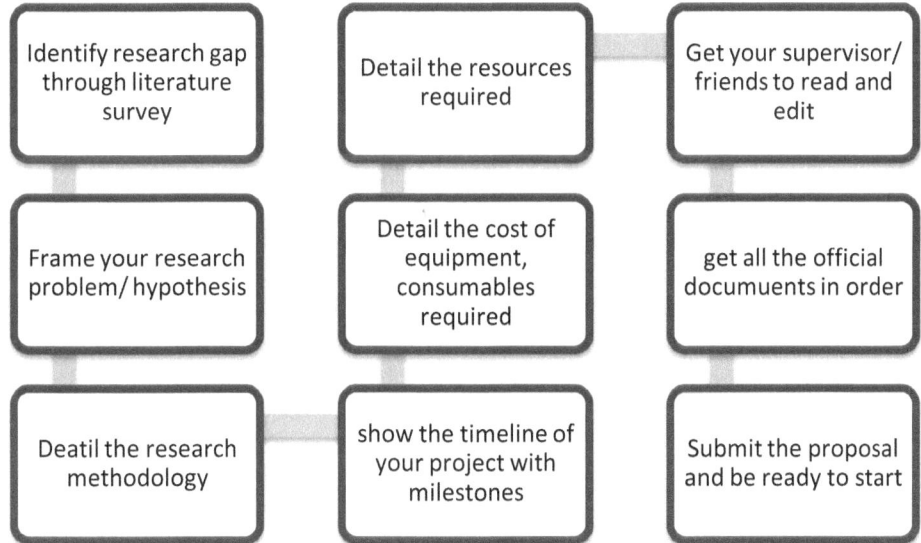

- o Place your research hypothesis in the research gap you have identified
- o Explain in detail the method you are going to adopt to gain answers for the questions framed in your hypothesis
- o Justify that the methods selected are the best available to gain the answers
- o Write your methodology by adhering to the codes and protocol established by learned societies and agencies in that field
- o Choose the right journal in your field, which is well regarded by eminent scientists/ faculty/ funding agencies as a reference
- o Detailed proposal – methodology you will adopt, details of experiments you propose to conduct
- o Detailed budget – contains listing of the cost of equipment/ software/chemicals/ consumables, which you require.

- o Detailed execution schedule of your project. Preferably, a bar chart that indicates what you plan to achieve in so many weeks/ months/ years.
- o Paperwork related to endorsement from your faculty advisor, institute, and other official certificates.

Most agencies funding student project might require the endorsement of a faculty advisor for the project. Therefore, it is advisable to maintain good rapport with the faculty working in your area of interest. Most faculty members will be readily willing to help in preparing the proposal and act as your project guide/ advisor if you show that you are serious about learning and executing the project in the right manner.

Inspiring Inventors

The man who invented flying ships 40.4

Ferdinand Von Zeppelin (1838-1917) was a German General and an airship manufacturer. Born in a noble family, he attended the polytechnic in Stuttgart and the military school to become an army officer. In 1858, he was made a lieutenant and was given leave to study science, engineering, and chemistry. However, he had to return to the army in 1859 for the Austro-Sardinian war.

In 1863, while he was in America as an observer in the American civil war, he made his first aerial ascent in a balloon made by John Steiner. This episode made him think about making an aircraft lighter than air. After his resignation from the army in 1891, he started serious efforts to design an airship and to secure funds for this project. After many setbacks, he gave a lecture on his airship design to the association of German engineers, who were impressed and made an appeal for his funding.

After securing funds from various quarters and using his property, he constructed Zeppelin airship LZ1 and it made its first flight over Lake Constance and remained airborne for 20 minutes. After many setbacks and improvements, airship LZ3 was accepted into army service and thus began a new era in air transportation.

His design was used to make many airships that were used to bomb Britain during the First World War. Airship Service also started between Europe and America and continued successfully until the Hindenburg disaster.

STEP 41

Private Funding - The Startup Route - You need just an Idea

Money is only a tool. It will take you wherever you wish,
but it will not replace you as the driver -Ayn Rand, Atlas Shrugged.

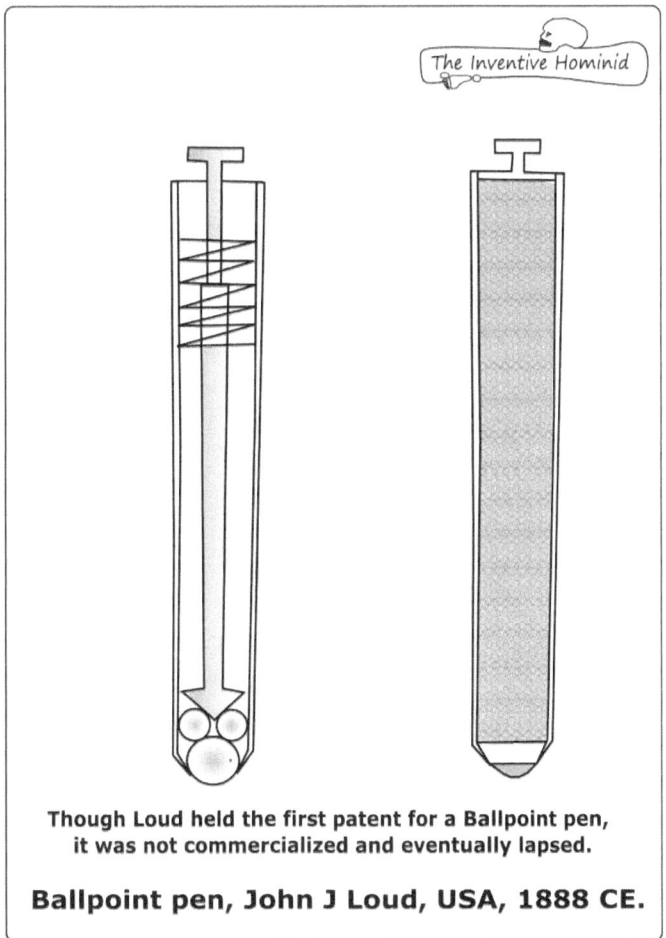

The Inventive Hominid

Though Loud held the first patent for a Ballpoint pen,
it was not commercialized and eventually lapsed.

Ballpoint pen, John J Loud, USA, 1888 CE.

TIH - Ballpoint Pen (41 IR 1, 41 TR 1)

Learning Objectives

After studying the contents of this chapter, you should be able to:
- Learn about the process of transforming your idea into a startup
- Understand the various stages of startup funding
- Learn about writing a business plan for your startup

Are you looking for private investment? Why not? Congratulations, welcome to the startup world.

If you think that your idea has the potential to grow into a big business, it is time to Startup. You can very well jump into the startup world if you think that your idea satisfies the following criteria:
- Solves a unique problem currently existing in the market
- Your solution is very effective when compared to existing ones
- Has the potential to grow quickly into a million dollar business

The typical cycle of a start-up 40.1, 40.2, 41.1

From start to growth into a big business is as follows:
- Brainstorm among friends and come up with an idea worth turning into a business
- Bootstrap the startup by gathering financial resources among friends and family
- Get the idea into shape by registering with an incubator, develop a prototype, and try for funding from seed investors or angel investors (investors who provide funding during the initial phase of a business in return for equity in the company)
- With seed funding, grow the company to break even phase (a time when the company has recovered the costs and will begin to make a profit)
- Now try for more funding to expand the business from venture capitalists who fund businesses in return for equity in the company
- Stabilize the business and plan an IPO (Initial Public Offering in which a company sells shares of its business to interested buyers for the first time in a stock market).
- Reap the profits and live happily ever after!

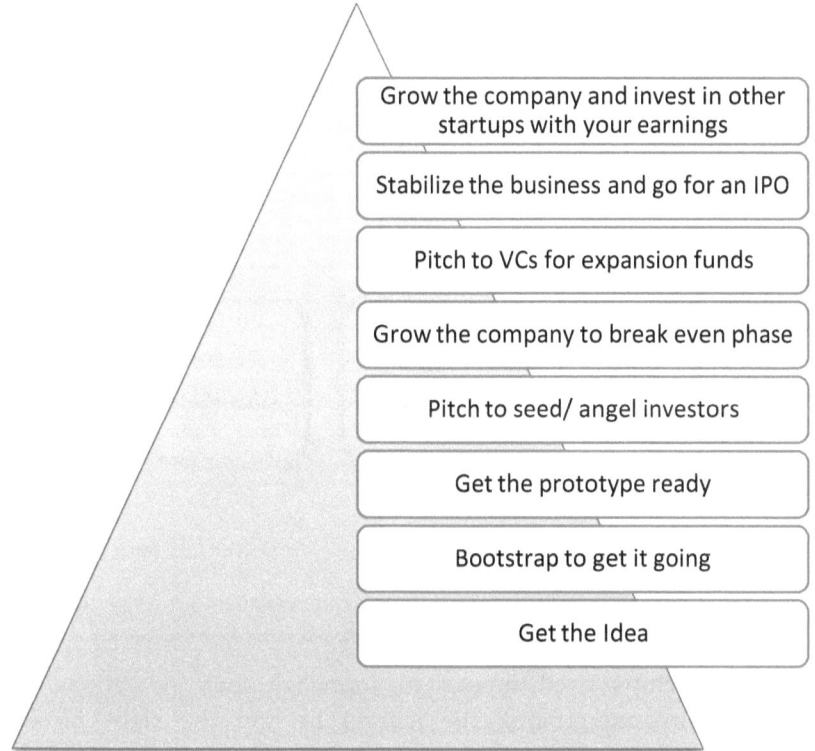

Flipkart, the Indian e-commerce aggregator was started by Sachin Bansal and Binny Bansal with an initial investment of INR 4,00,000 (USD 5600). In 2015, the two co-founders debuted on the Forbes India rich list with a net worth of USD 1.3 Billion each. In addition, not to forget that Facebook started as a website co-created by Mark Zuckerberg, then a second year student at Harvard University.

Stages of Startup Funding/ Financing

I - PRE-SEED CAPITAL

This is the first stage where you want to transform your idea into a prototype. At this stage, there are three options for funding.

FFF (Fools Friends and Family) - you can convince people who know you, like friends and family who can fund you based on the trust they have in you and not your idea.

Startup Angels - mostly successful startup founders who want to help other startups in their early stage.

Incubators/ Accelerators - Govt. or private organizations who provide much needed capital for 5-10% of equity (ownership of the company).

I -Pre-Seed Capital Idea to prototype (FFF, Angels, Accelerators)	II - Seed Capital Formal operations (super angels, Crowd funding)	III - Growth Stage Expansion Stage (VCs - Series A, B, C)

II - SEED CAPITAL

At this stage, startups need money to formalize their operations like registering a company and testing the market to find the right product specification. Funding at this stage can be had from:

Angel investors, super angels, and early stage venture capital funds.

Crowd funding: where startups are funded by a large number of people in return for some reward such as discounted products or some equity in case of equity crowd funding.

By raising small amounts from a large group of people, people have been able to fund their projects this way. It is estimated that billions of dollars are being raised through crowd funding for a variety of reasons.

Funds are raised through an online platform where people can create an account and promote their projects.

People raise funds for purposes ranging from building innovative projects, medical treatment, charity, etc. If you believe that you have an innovative product idea and are not able to raise funds from conventional sources, you can try crowd funding. Individuals have raised millions of dollars by pledging contributors a reduced price for the developed product or exclusive deals on them.

Popular crowd funding sites include Kickstarter, Indiegogo, Fundanything, Companisto, etc.

III GROWTH STAGE

Series A Round - Traditional Venture Capital firms provide funding at this stage in return for 15 - 30% of equity. At this stage, the startups usually have a successful product and they need capital to improve their business process.

Series B - At this stage, the startups have an established user base and want to scale-up their operations and obtain funding from VC firms.

Series C and ensuing rounds - At this stage due to their mature business model, they are able to obtain funding from private equity firms and investment banks. At this stage, most startups plan their IPO (Initial Public Offering in terms of shares in the stock market). Some startups might be acquired by big companies for a large amount of money making the startup founders multi-millionaires.

How to write your startup business plan to attract investors?

ELEMENTS OF A STARTUP BUSINESS PLAN 41.2

- o   What do you want to sell – is it a product or a service?
- o   What is the need or gap, which it is going to fulfill?
- o   Is your idea going to satisfy a need or luxury?
- o   What are the current alternatives or competitors of your idea?
- o   What is the USP (unique selling point) of your idea, which will differentiate it in the market?
- o   How big is the market for your idea?
- o   How much are the people willing to pay for your product/ service?
- o   Can you scale it up across the city/ state/ country/ world?
- o   Have you checked the patent database to ensure that you are not infringing on patented ideas?
- o   Do you have a prototype ready?
- o   Do you have a customer base?
- o   How much money is required to release the first batch into the market?
- o   How much money is required for manufacturing, marketing, workforce, legal requirements?

1 •what do you want to sell? is it a product or a service?

2 •have you checked the patent database? is it similar to anything existing?

3 •what is the need for it? what gap is it going to fill?

4 •is your idea going to satisfy a need or a want

5 •what are the alternatives or competitors currently in the market

6 •what is the USP (unique selling point) of your idea or why people will prefer your product/ service over others

7 •how big is the market? which is the demographic you are targetting?

8 •how much are people willing to pay? did you sound/ survey the market?can you scale it up across the city/ nation/ world?

10 •do you a protype ready? can we release the first batch of products?

11 •how big is the customer base? what are the costs involved in manufacturing/marketing/ logistics/ servicing?

12 •who will be the people involved in the company and what are their roles? how many do we have to recruit?

13 •when will the company break even and genarte profits? what is the plan for the next 1, 5, 10 years?

- Who are the people involved in running the startup and what are their roles and stakes?
- When will the company break even and start generating profits?

o   What is the plan for the next 1, 5, and 10 years?

Just prepare a concise presentation addressing the above points and there you have a detailed winning startup business plan.

Inspiring Inventors

The Navy man who tamed the air waves with a series of startups

Arogyasami Paulraj (born 1944) is an Indian-American Electrical Engineer and Emeritus Professor at Stanford University. Paulraj born in Pollachi, India, joined the Indian Navy at 15, and served for 26 years. While at Indian Navy, Paulraj graduated in Electrical Engineering from Naval College of Engineering, Lonavla and obtained his Ph.D. in Electrical Engineering from IIT, New Delhi. Among his contributions while in the Indian Navy are improved trans-receiver display for Sonar 170B, large surface ship sonar APSOH, establishment of three national level research centres: Centre for AI & Robotics, DRDO; Central Research lab, Bharat Electronics and CDAC, DoE (as co-founder). Until today, due to his contributions, Indian Navy is self-reliant for Sonar technology [41.3].

Yearning for a highly stimulating research environment, he wrote to Prof. Kailath of Stanford to join his Lab. Rejected twice, he wrote to the Professor for a third time asking him to just give him a desk at his lab without any salary or stipend. He went there in 1984 and came up with the ESPRIT. Back in India in 1986, he started various R&D Labs. He joined Stanford University again in 1991, and in 1992 invented the MIMO (multiple antennas at both ends of a wireless link). This technology lies at the heart of current high speed Wi-Fi and 4G mobile phones. Further, he has pioneered advanced technology startups, which were acquired by big companies such as Lopsan Wireless for MIMO-OFDMA, Beceem Communications for WiMAX chips and Rasa Networks for using AI tools in Wi-Fi networks. Paulraj is the co-inventor of over 80 patents and the recipient of Padma Bhushan in 2012. He has received various International awards and got into the USPTO National Inventors Hall of Fame in 2018 [41.4].

STEP 42

TOP COMPANIES THAT ORIGINATED AS STARTUPS BY COLLEGE STUDENTS

Startups don't win by attacking. They win by transcending. There are exceptions of course, but usually the way to win is to race ahead, not to stop and fight
- Paul Graham, "Mean People Fail"

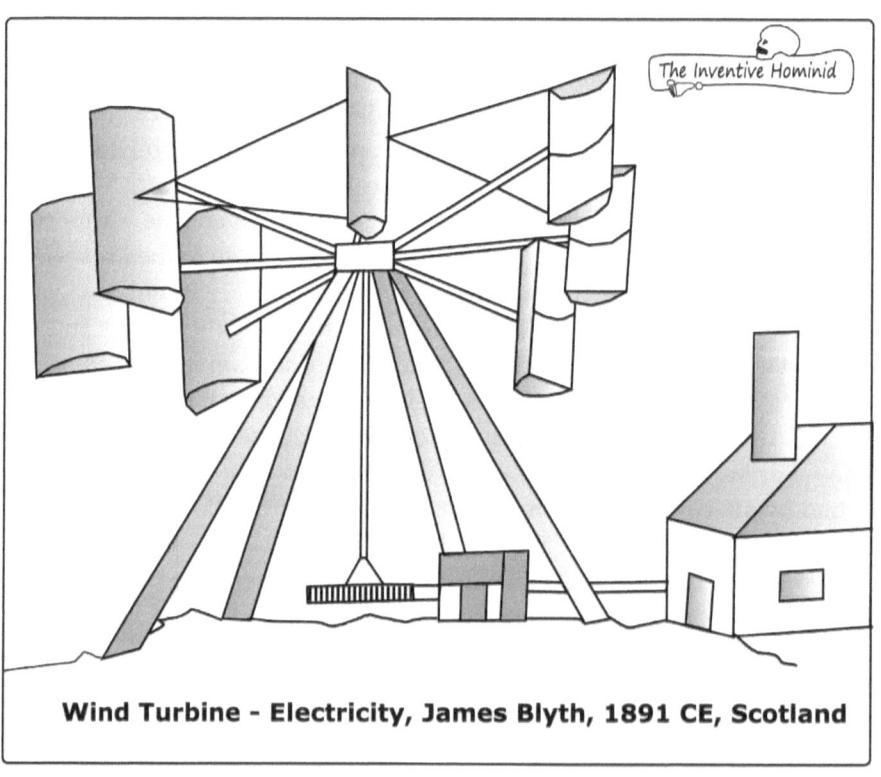

Wind Turbine - Electricity, James Blyth, 1891 CE, Scotland

TIH - Wind Turbine (42 IR 1, 42 TR 1)

Learning Objectives

After studying the contents of this chapter, you should be able to:

- Know about the top companies, which originated in college
- Learn about the global startup scenario and ranking of countries and cities
- Understand the attributes of successful startup founders

Successful startups by college students 42.1

College life is the period, when exuberant students blossom into thinkers and doers. Many a successful company today originated in the minds of young college students like:

- Dell
- Google
- Facebook
- Dropbox
- Microsoft
- Napster
- Reddit
- Snapchat
- Wordpress
- Yahoo

So build a team of likeminded friends and feel the synergy rush through the team to accomplish great things. This is the time to ideate and start-up. So why wait? You can read more about great college startups at fastweb.com [42.1].

Global Startup Scenario

As per the global startup ecosystem report (GSER, 2018) [42.2], the top four growing subsectors in the Startup Industry are:

- Advanced manufacturing and Robotics which has seen a 189% increase over 5-years in early stage funding deals
- Agricultural technologies (Agtech) & New Food technologies (171% increase over 5 years)
- Blockchain technologies (163% increase over 5 years)
- Artificial Intelligence (AI), Big Data and Analytics (77.5% increase over 5 years)

The other startup sub-sectors that have powered growth over these years are:

- Advertising technologies (Adtech)

- o Gaming
- o Digital Media
- o Financial technologies (Fintech)
- o Cybersecurity
- o Educational technologies (Edtech)
- o Clean/pollution free technologies (Cleantech)
- o Health and Life Sciences technologies
- o Biotechnology

Global Startup Ecosystem ranking

Countries around the world have recognized the potential of Startups and have initiated a slew of measures to encourage homegrown startups. They are also keen to attract talent from around the world to setup base in their country. In fact, there is a big competition among cities within countries to emerge as the top ranked destination for startups.

Country ranking for best startup ecosystem 2018 [42.3]

- o United States
- o United Kingdom
- o Canada
- o Israel
- o Germany
- o Sweden
- o Denmark
- o Switzerland
- o France
- o Singapore

However, other countries are providing a good ecosystem for startups in specific sub-sectors such as China for Advanced Manufacturing, India for Education Technologies, etc.

Cities ranking for best startup ecosystems 2018 [42.3]

- o San Francisco (Silicon Valley), US
- o New York, US
- o London, UK
- o Los Angeles, US
- o Berlin, Germany
- o Boston, US

- ○ Tel Aviv, Israel
- ○ Chicago, US
- ○ Seattle, US
- ○ Paris, France

Other prominent cities around the world are fast catching up, boosted by special incentives and work force availability.

The key qualities/ attributes found in successful startup founders 42.4, 42.5, 42.6, 42.7

- ✓ Willingness to learn/ experiment
- ✓ Leadership skills
- ✓ Passionate about their goals
- ✓ High Emotional Intelligence/ empathy
- ✓ Open to criticism/ feedback
- ✓ Quick learning
- ✓ A bit crazy/ lateral/ out of the box thinking
- ✓ Excellent knowledge of Technology
- ✓ Patience/ ability to focus long and hard
- ✓ Energy/ enthusiasm
- ✓ Vision/ ability to see the long-term big picture
- ✓ Resilient against all obstacles
- ✓ Serial innovators/ innovation as a way of life
- ✓ Good communication skills
- ✓ Willing to start from scratch
- ✓ Willing to accept failure and start again from scratch
- ✓ Focus on working prototype over capital
- ✓ Willingness to do the hard spadework by themselves/ getting the hands dirty
- ✓ Self-motivated/ self-starters
- ✓ Decisive in thinking
- ✓ Resourceful/ inventive
- ✓ Consider customer as King/ respect the customer
- ✓ Meticulous planning/ micro-level/ attention to detail

Are you overwhelmed by the long list of attributes? You just have to remember that all these attributes can be cultivated by anyone with a bit of determination and focus. Therefore, starting from today, just believe that you have all these attributes and act like one. Take a printout and paste it in your mirror, read every day whenever you pass it. Hammer it into your brain and let your sub-conscious mind take over. It is just a matter of time and you will be surprised by the positive results in your quest.

Inspiring Inventors

The inventor who worked as a waiter to buy a motorcycle
42.8, 42.9

G Doraiswamy (1893-1974) is an Indian inventor referred to as the Edison of India. His contributions span the fields of electrical, mechanical, agriculture and automobiles. Doraiswamy lost his mother when he was a few months old baby. A few years later, after a failed attempt to send him to school, he was sent to stay and manage his father's farm. Here he developed the habit of reading various books to improve his knowledge. Upon seeing a British officer riding a motorcycle in his village, he dreamt of owning one. To pursue his dream, he went to Coimbatore and took up the job of a waiter in a hotel. After working for three years in a hotel, he had saved about 400 rupees. With this money, he located the British officer and convinced him to sell his motorcycle. Doraiswamy rode the motorcycle around with pride and became skilled with its repair by taking it apart and putting it back again numerous times.

Later, he met the Britisher Robert Stanes who was running a motor transport company, seeking the job of a mechanic. Stanes, impressed by Doraiswamy's work ethic, loaned him a motorbus and asked him to repay in small amounts. Thus, starting with one bus in 1920, Doraiswamy became the owner of 70 buses in 10 years.

In 1936, his ultra sharp shaving razor won third prize in an international industrial exhibition in Leipzig, Germany. Not content with his successful transport business, he established National Electric Works in 1939 and manufactured the first electric motor in India. Some of his indigenous inventions are Refrigerator, Radio, Clock, Juicer, vote recording machine, kerosene run fan, mechanical calculator, 2-seater petrol car, hybrid varieties of Cotton, Maize, Papaya, banana, coconut, etc. Many of his inventions were not given adequate recognition by the British Government and could not be commercialized. Unfazed, he setup many industries and industrial training institutes, polytechnic colleges and engineering college. These industries and institutions continue to serve the people to this day.

STEP 43

MAINTAINING SECRECY – INDUSTRIAL ESPIONAGE AND FRAUDSTERS

In making tactical dispositions, the highest pitch you can attain is to conceal them -
Sun Tzu, Art of war.

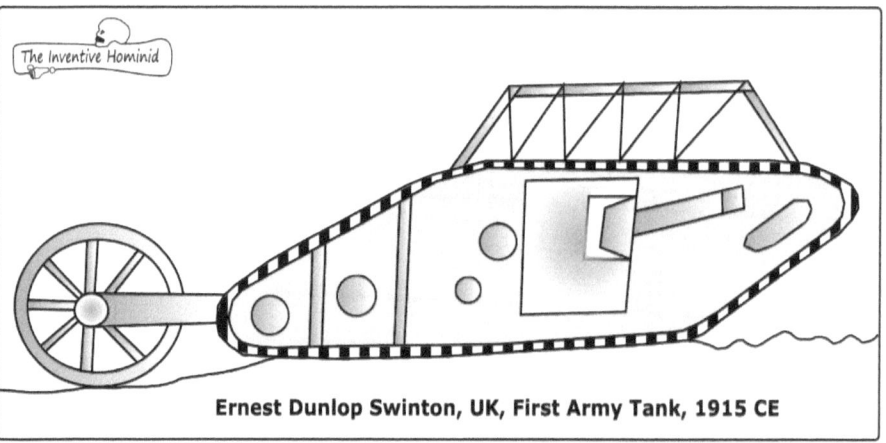

The Inventive Hominid

Ernest Dunlop Swinton, UK, First Army Tank, 1915 CE

TIH - Army Tank (43 IR 1, 43 TR 1)

Learning Objectives

After studying the contents of this chapter, you should be able to:
- Learn about the dangers of your idea being stolen
- Understand ways to protect your idea/ design/ product

Remember, it is not enough if your project is good; it has to be better than the rest of the projects done by your peers.

One important aspect of delivering a good project is to guard your ideas from prying eyes and ears. It is a 'dog eat dog world' out there. Keep your central idea and the innovative component a closely guarded secret among your group. Watch out for sweet talking people trying to pry out ideas from your mouth.

Even after, you obtain a patent for your Intellectual property; it is important that the core knowledge about your invention be protected from industrial espionage. In this day of digital hacking, many a company has lost its ideas to competitors and eventually forced out of the market due to a flood of cheap imitations.

WAYS TO PROTECT YOUR IDEA/ INVENTION

- ✓ Document everything: document your ideas, progress of your work, and store it in a secure manner. Store documents with passwords and better use encryption software.
- ✓ Patent it: file a patent and get protection in a legal manner
- ✓ Get help: get professional help by hiring a patent consultant once you figure that your idea is worth a lot.
- ✓ Chose your partners carefully: chose partners/ collaborators carefully, and make sure they sign a Non-disclosure agreement.
- ✓ Do not reveal too much: while pitching to potential investors, reveal only what is necessary to convey the idea with as little information as possible.
- ✓ Trademark your name and your unique design.

Hence, if you have a potential million-dollar idea, patent it with professional help and secure your data with expert state-of-the art security systems.

Inspiring Inventors

The wagon driver who made skyscrapers possible
43.3, 43.4

Elisha Graves Otis (1811-1861) was an American inventor known for inventing a safety device, which prevents elevators from falling when hoisting cables fail.

Otis was born in Halifax, moved away from home, and lived for five years working as a wagon driver. After marriage, he moved his family to Vermont hills and built a gristmill. Not able to earn enough, he converted it into a sawmill, which also did not attract customers.

Then he started building wagons and carriages. After the death of his wife, he moved around working in various jobs and finally started working as a manager in a factory. While wondering how to move all the things to the upper levels of the factory, he thought of designing a hoisting platform with a safety device to prevent it from falling. He built one and made several sales, but could not attract large orders.

His fortunes changed, when at the New York World fair in 1853, he demonstrated his safety elevator to the public. While standing on a hoisted platform, he ordered for the ropes holding it to be cut. As the public gasped at the danger, the platform did not fall, but came to a stop within inches of moving down owing to its safety mechanism.

This public demo allayed safety fears and his elevator came to be used in all skyscrapers, enabling them to be constructed ever higher. Today the company he founded, the OTIS Elevator Company functions around the world with revenues of $12.34 billion (2017).

STEP 44

IDEA TO PROTOTYPE TO PRODUCT

One should start acting after planning about everything,
Will be a disgrace to procrastinate after the action has started.
-Thiruvalluvar, Thirukkural, 467 – Decision making

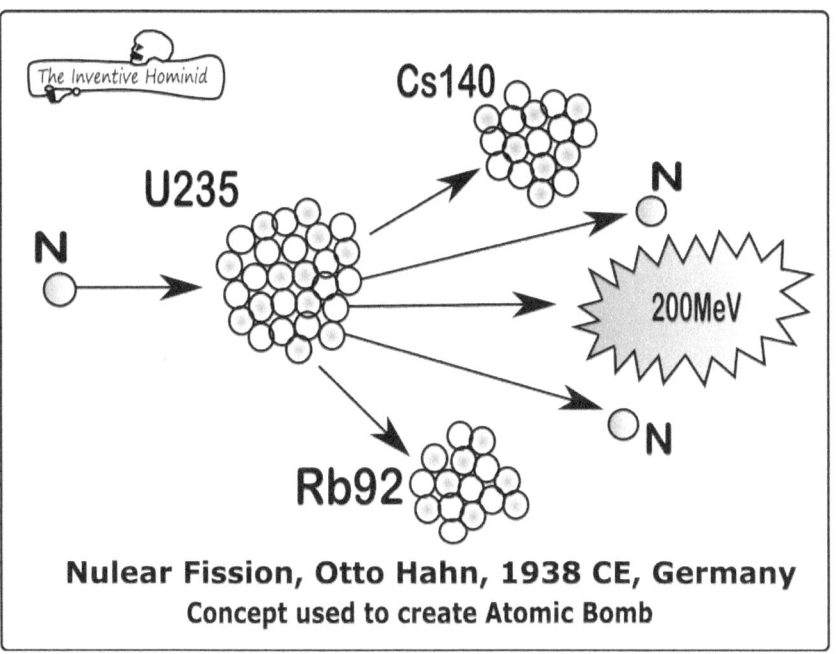

Nulear Fission, Otto Hahn, 1938 CE, Germany
Concept used to create Atomic Bomb

TIH - Nuclear Fission (44 IR 1, 44 TR 1)

Learning Objectives

After studying the contents of this chapter, you should be able to:
- Learn about rapid prototyping through 3D printing and companies offering assistance through websites
- Learn about companies offering online resources to design and print your circuit boards
- Understand the process of prototyping software and tools available

You have a dream innovation and want to create a prototype. A prototype is necessary to pitch it to potential investors and to get feedback from multiple sources to improve the final product. There are many efficient ways to create your first prototype and these are discussed below:

RAPID PROTOTYPING THROUGH 3D PRINTING 44.1, 44.2

Steps to turn your idea into a working machine/ component:
- ✓ Sketch your idea on paper and finalize the dimensions of components as per design calculations
- ✓ Turn your sketches into 3D drawing using CAD software
- ✓ Using the 3D CAD drawings, approach the nearest available 3D printing facility to print out the components within hours
- ✓ If you are unable to access 3D printing facility, use the online services to 3D print your components like 3Ding, Think3D, 3Digiprints, Garuda3D, Shapeways, Imaterialise, Sculpteo, Grabcad, Thingiverse, Makexyz, Ponoko, You3dit, etc.
- ✓ Once you send your 3D CAD files to them, these online services companies will print out your components and ship it to you for a fee.

Companies like SKANECT offer low cost scanning devices, which you can use to 3D scan objects, if you have made physical models of components, and send it for 3D printing.

Best 3D printing sites  Search Alert

Easy is it not. So let your imaginations run wild and do not be limited by available components. Now you can just create any component from your dreams. Draw, mail, 3D print, and use.

Though 3D printing is rapidly catching on, still we must not forget the traditional prototyping through lathes and CNC machines. They are also widely available and used to date.

CIRCUIT BOARDS 44.3

Steps to create circuit boards:
- o Design the circuit board for your application by selecting the appropriate components as per design specifications
- o Download an Electronic Design Automation (EDA) Tool to design your circuit
- o Get the components, assemble your circuit board, and test your application
- o You can also download free PCB design software or use licensed software to design a Printed Circuit Board (PCB)
- o After checking the correctness of your PCB design, extract the bill of quantities (list of components)
- o Buy the components or order online
- o Get your PCB printed or alternatively, get your PCBs printed and delivered to you by ordering online from PCB Power, Lion Circuits, PrestoPCB, Oshpark, etc.
- o Alternatively, you can create your own PCB at home using inexpensive materials by following DIY (Do It Yourself) videos online.
- o Assemble the components and run your application

 **Best websites for**
PCB design and ordering

SOFTWARE 44.4

Creating a prototype of software helps visualize the final version in a better manner by the creator and the client. It also helps finalize the input and output specifications along with how the software will look like (User Interface). A prototype will help avoid costly modifications after the entire software is built. It will also help the creators to try out different versions of the user interface and get feedback from users to arrive at a more effective final version.

The various software tools available for creating prototypes are [44.5]:

o   Adobe XD
o   UXPin
o   Origami Studio
o   Marvel
o   Proto.io
o   Pidoco
o   Atomic
o   Invision
o   Framer

Depending upon your requirement of a prototype (web or mobile app) you can just browse around and find the right set of tools easily, to speed up the development of a prototype and the final version.

Top software tools for prototyping

𝒥𝓃𝓈𝓅𝒾𝓇𝒾𝓃𝑔 𝒥𝓃𝓋𝑒𝓃𝓉𝑜𝓇𝓈

The windows maker who wanted to be a lawyer
44.6, 44.7, 44.8

William Henry Gates III or Bill Gates (born 1955) is the principal founder of Microsoft Corporation. Bill Gates born in a family of high achievers grew up in an environment, where winning was rewarded and failure was punished. At school, Gates showed more in interest in programming and was allowed to pursue his passion. He pursued his interest in programming by learning languages like Fortran, Lisp and machine language. In School, he wrote a program to schedule student classes and wrote a payroll program in COBOL for Information Sciences Inc. As a school kid, Gates was always reading something; his parents banned him from bringing his book to the dinner table. At 17, Gates along with Paul Allen made automatic traffic counters with Intel 8008 processor and called it Traf-O-Data.

In 1973, Gates scored 1590/1600 in college entrance SAT tests and enrolled at Harvard College. There he chose a pre-law major course, but also took mathematics and computer science courses. After the release of MITS Altair 8800 in 1975, Gates and Allen jumped at the opportunity to start their own Software Company.

Gates contacted MITS and informed them that they were working on a BASIC Interpreter for Altair 8800. He did this to gauge their interest, as at that time they were not working on an interpreter. When MITS agreed to meet them and called for a demo, they sat and developed the BASIC Interpreter. The demo was a success, Gates and Allen named their company Microsoft, and the rest is popular history. Gates instead of selling the software, stuck a franchise deal with computer manufactures, which gave them payment for each computer sold. Gates grew Microsoft into a billion dollar company and became the richest man in the world for many years.

STEP 45

PUBLISHING YOUR RESEARCH PROJECT RESULTS IN JOURNALS

......do not be impressed by the imprint of a famous publishing house or the volumes of an author's publications. Bear in mind that Einstein needed only seventeen pages for his contribution that revolutionized physics -Andreski, Social Sciences as Sorcery

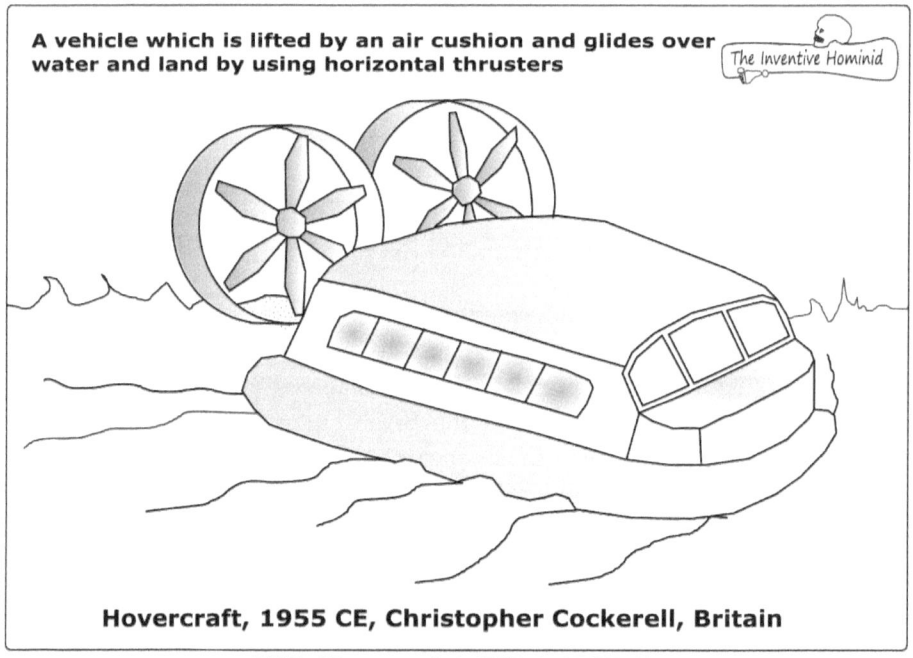

A vehicle which is lifted by an air cushion and glides over water and land by using horizontal thrusters

The Inventive Hominid

Hovercraft, 1955 CE, Christopher Cockerell, Britain

TIH - Hovercraft (45 IR 1, 45 TR 1)

Learning Objectives

After studying the contents of this chapter, you should be able to:
- Know about research journals, journal indices, impact factors
- Understand the steps in writing a research paper
- Importance of avoiding plagiarism and using tools for referencing

WHAT ARE JOURNALS - HOW TO IDENTIFY THEIR QUALITY?

Journals are scientific publications that publish the research work of scientists. Reputed journals have a good editorial board consisting of reputed academics/scientists in their chosen area of research. Research papers received by these journals are sent to experts for review of the results or the theory reported in the paper. Only after a thorough review by a number of experts, a paper is accepted for publication.

As for the quality of the journal is concerned, it is better to consult researchers in that particular field and look at their publications, which you can do it through platforms like Google Scholar, ResearchGate, Academia, Etc.

SCIENCE CITATION INDEX (SCI) 45.1

SCI is a list of significant journals selected through rigorous criteria in all areas of academics and research. SCI expanded is a much larger list of about 8500 journals. These indices serve as a ready reckoner to find prestigious journals in your field. Another widely used Journal list is that of SCOPUS, which covers nearly 22,000 journals and provides a searchable database also containing detailed information about the authors [45.2].

CITATION

Citation is a reference to a published or unpublished source. When you build upon previous work, you have to acknowledge the author of that work by giving information about the author/s, title of their work, publication details like journal name, issue number, date, etc. This is called a citation. When more number of people cite your work, you are considered to have produced a work with a big impact or more citations.

IMPACT FACTOR

An impact factor is the measure of a journal's reputation, as it measures the number of times its papers are cited (taken as a reference for further research) by researchers around the world.

WHAT ARE THEIR IDEOLOGY/ PHILOSOPHY?

To identify a journal fit for your research paper, you can start by browsing through the list of journals preferred by top researchers in your field. Top researchers can be identified from the faculty/ scientist profiles displayed on the websites of departments in leading institutes of the world. You can also lookup the h-index of researchers from Google Scholar, Scopus. H-index is a measure of consistent performance of a researcher in publishing papers with maximum citations.

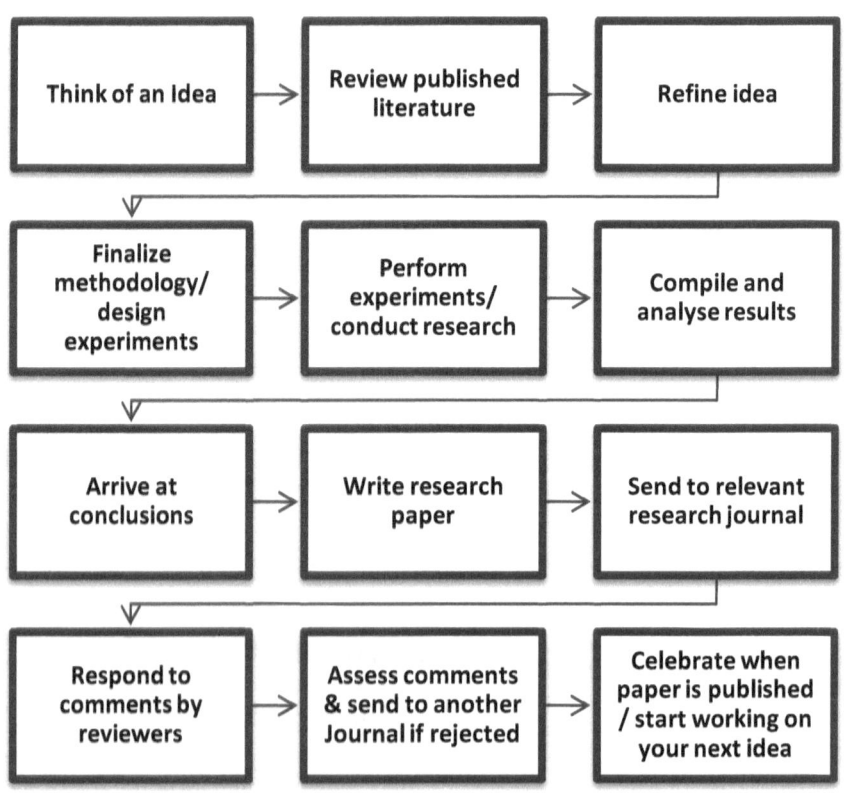

Once you identify a set of journals, then you can browse through their publications from their website, read their aims and objectives, their research focus and people on their editorial board. This way you can understand the ideology or philosophy or the core area of research a particular journal stands for. The entire process of publishing a scientific paper is given in the flow chart.

JOURNAL WRITING STYLE

Every journal has a specific requirement in terms of formatting and writing style. It can be known from the section usually titled 'information for authors. It will specify the acceptable word processing software, document formats, image quality required, image file formats, format of references, etc.

WRITING- REVISING-REWRITING-REVISING

Communicating your results through a research paper is an art you will perfect through the various comments and rejections from reviewers and editors. A journal paper has to be concise and at the same time explain your findings in a clear unambiguous manner. To arrive at your final paper, you might have to write down your thoughts and revise it a number of times. Be open minded to accept criticism and comments from friends and your faculty supervisor and incorporate them, if it improves the paper.

AVOIDING PLAGIARISM 45.3

What is plagiarism? If you use ideas, words, figures, etc. from papers, reports, websites, books without acknowledging them, then it constitutes plagiarism.

How to avoid plagiarism in your report?

o   If you want to use the same exact words/ phrases, use it within quotes and cite the source by giving full reference as per accepted convention including page number of the source.
o   Paraphrase / summarize the idea from others without using their words and cite them properly.
o   Obtain prior permission in writing for using tables, figures, etc. and cite them properly.
o   Use good plagiarism checker software before sending your paper for publication.

USING AUTOMATED REFERENCE SOFTWARE 45.4

A major time consuming factor in writing a paper is citing other papers as references and arranging them as per the format specified by the journal. A number of tools are available to make the job of a researcher easier in this aspect. Some of the tools available are Zotero, Mendeley, Refworks, Endnote, Citeavi, bibme, etc.

Most of these sites offer free versions that are sufficient for beginners. Using these tools will free up a lot of time, which can be spent on research and analysis, rather than on mundane referencing.

𝓘𝓷𝓼𝓹𝓲𝓻𝓲𝓷𝓰 𝓘𝓷𝓿𝓮𝓷𝓽𝓸𝓻𝓼

The Man called the 'Mayor of Silicon Valley' 45.5, 45.6

Robert Norton Noyce (1927-1990) was an American behind iconic companies such as Fairchild Semiconductor and Intel Corporation. He is also credited as the co-inventor of the first integrated circuit or microprocessor that kick started the PC revolution and gave Silicon Valley its name. With a Rhodes Scholar for his father, Noyce excelled in mathematics and science and took a college level physics course while still in school. Though excelling at college, he was suspended for one semester for stealing a pig and roasting it.

After obtaining his PhD from MIT, he co-founded Fairchild Semiconductor, before leaving to start Intel Corporation with Gordon Moore in 1968. He declined lavish facilities and benefits at Intel, and encouraged a relaxed atmosphere to foster creativity. He received many awards and was credited in 15 patents.

STEP 46

STUCK IN A RUT: WHAT TO DO WHEN YOU HIT A ROADBLOCK IN YOUR PROJECT

What does not kill him, makes him stronger.
- Friedrich Nietzsche, Ecce Homo (1888).

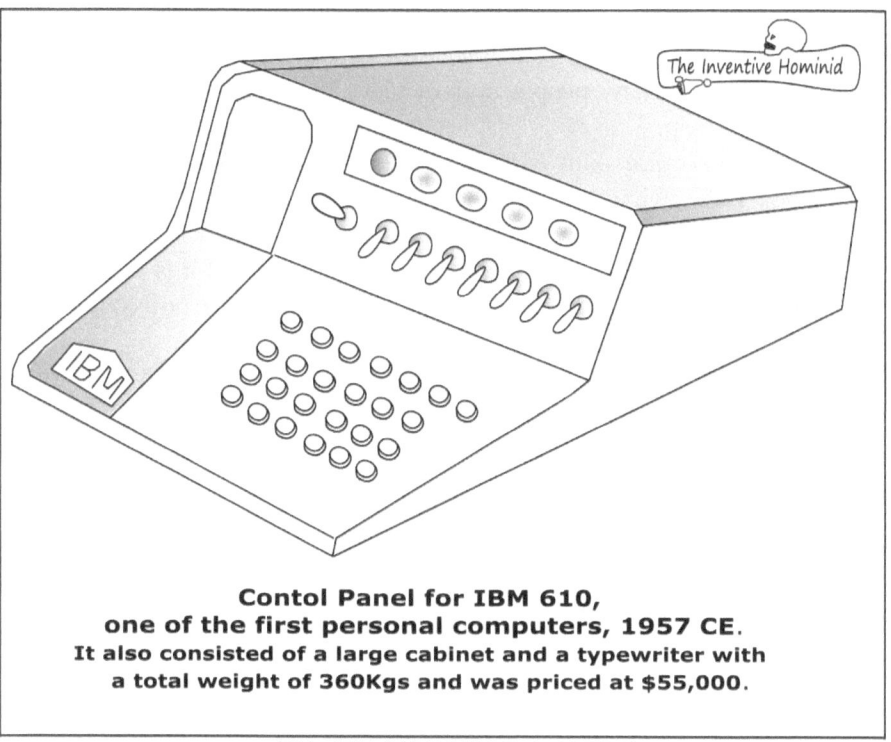

**Contol Panel for IBM 610,
one of the first personal computers, 1957 CE.**
It also consisted of a large cabinet and a typewriter with
a total weight of 360Kgs and was priced at $55,000.

TIH - IBM 610 PC (46 IR 1, 46 TR 1)

Learning Objectives

After studying the contents of this chapter, you should be able to:
- Understand the importance of hanging on when your project stalls
- Learn about ways to refresh yourself and move your project

Projects aimed at innovation in any field have to overcome the many problems that arise before reaching the stated objectives. History of invention is replete with stories of inventors overcoming a great many obstacles, which distinguishes them from mere mortals. So what do you do when you hit a seemingly dead end in your project?

STEPS TO OVERCOME OBSTACLES

- Reassess your goals and objectives – are they relevant and feasible
- Ask for help – get on Quora, LinkedIn groups, Google groups, your uncle at a company, your distant relative at a research lab or just your friend, too many people underachieve because they were too shy to ask for help.
- People acquire skills and knowledge over a considerable period and you know what, they will be more than happy to share it with you. You just have to ask sincerely.
- Take a break from your project for a day or two, if it is teamwork, let everybody take a break and indulge in anything not connected to academics.
- Regroup after the break and you will find that everybody is energized and a lot more creative in overcoming the obstacle.
- Most of the times when we feel hopeless about achieving something, usually it is not about our goals or abilities, it simply means your mind and body needs a break.
- Schedule regular breaks from your intense project activity and feel the amazing energy and rejuvenation of your mind.
- If your project activity involves physical effort, then your body needs a rest. If your project is about sitting long hours in front of a computer, then what you need is an intense physical activity like walking, jogging, games, etc.

Finally, we should understand that completing a project is an academic requirement and not a life-threatening situation. Reassess the objectives and reframe them to complete the project.

At times, it is important to lose a battle, in order to win the war.

Buck up, maybe you chose a problem, which needs a larger time frame to solve it. You can work on the problem again after you put the academic requirements aside.

Inspiring Inventors

He dropped out of college, but never stopped engineering
46.1, 46.2

Uddhab Bharali (born 1962) is an Indian inventor from Assam who has more than 100 inventions to his credit. Bharali was often made to stand out of his class, as he was always asking difficult questions to his teachers. He was often promoted to higher classes; thereby finishing his schooling aged 14. He joined mechanical engineering, but was unable to finish the course due to financial difficulties. Staring at a huge debt his family had incurred, he decided that the only way to repay would be to innovate and sell cheaper machines. He decided to sell polythene covers to Tea Estates and instead of buying a machine for INR 400,000 (USD 5600); he built the same machine himself for INR 67,000 (USD 950).

Always under pressure from the banks, he went on an invention spree, creating peeling machines for betel nut, cassava, garlic, jatropha, safed museli; redesigned paddy grinder, bamboo-processing machine, cutter for green tea leaves, etc. He came to the limelight, when his pomegranate de-seeder machine won a competition by NASA and acknowledged as the best in the world. Since 2005, the National Innovation Foundation has supported him. He further turned his attention towards innovation for the benefit of differently-abled people with a variety of machines and devices. He now takes up any problem, which is given by the NGOs and Government organizations for the benefit of the society. He single handedly supports various families with disabled people, trains unemployed youth, and plans to build an old age home. He was awarded with various honors by the government and NGOs. He was awarded an honorary doctorate by Assam Agricultural University and continues his invention spree until date.

STEP 47

WHAT IF, YOU FAIL IN YOUR PROJECT GOALS?

"When we tackle obstacles, we find hidden reserves of courage and resilience we did not know we had. And it is only when we are faced with failure do we realize that these resources were always there within us. We only need to find them and move on with our lives". -A. P. J. Abdul Kalam

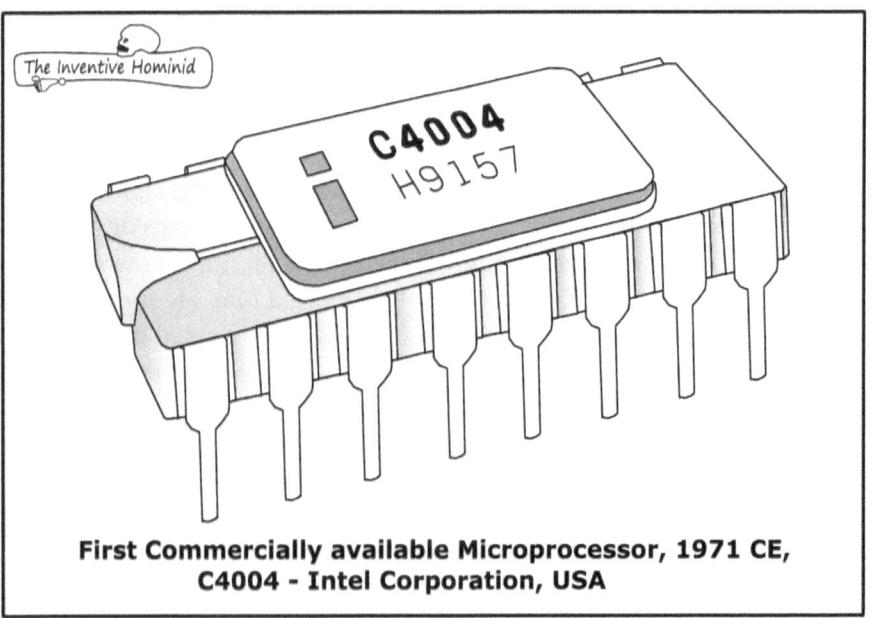

First Commercially available Microprocessor, 1971 CE, C4004 - Intel Corporation, USA

TIH - Intel C4004 (47 IR 1, 47 TR 1)

Learning Objectives

After studying the contents of this chapter, you should be able to:
- Understand that failures are never permanent
- Learn that a project is just an academic requirement to demonstrate your learning

THERE IS NO SUCH THING AS FAILURE IN THIS WORLD

We can maybe term them as temporary setbacks. History is full of happenings where a person has lost it all, his family, property or even his kingdom. Lo behold, after a few years of single-minded pursuit, he/she gets it all back.

Think about Steve Jobs who was booted out of Apple, the company he nurtured. How he got back to Apple is the stuff of Techlore. Whether Jobs left after the failure of the Macintosh or was he pushed out might be a point to be settled. Nevertheless, the fact remains that Jobs got back to lead Apple to greater heights after what can be termed as a failure or more appropriately temporary setback.

Now coming to your project, there is no such thing as a failed project. A project is an activity where you demonstrate your skills to combine knowledge from various engineering courses that you have taken to come up with a solution for an engineering problem. Phew!

Therefore, what matters is whether your approach to solve a problem has demonstrated your organized way of thinking. That is all. The actual success of a solution is not critical. Rather, it shows that instead of sticking to a traditional non-risk approach of selecting a safe problem. You have taken a risk of trying to solve a bigger problem. Kudos to your effort!

It also means that in the coming months, if you tackle the problem from a different angle or approach, you will be able to solve the problem. Many great inventions were not achieved over a period of weeks or months, the time given for an academic project, but rather over the period of many years of tenacious determined effort.

So get up, dust yourselves and keep your confidence levels high and document your work in detail about how your work can be extended in the future with different approaches and submit the report.

Inspiring Inventors

Success after 5127 attempts/ prototypes 47.1, 47.2, 47.3

James Dyson (born 1947) is the British inventor of the bag less vacuum cleaner. After schooling, he studied at the Byam Shaw school of Art (1965-66), and then at Royal School of Art studied furniture and interior design (1966-70). Determined to create a better vacuum cleaner than the existing one that clogged with dust and reduced the suction, he started working on his idea. After being kicked out from his company, he continued his work while his family was supported partly by his wife's salary. After 5 years of determined hard work and 5127 prototypes, he created a bag less vacuum cleaner that worked. After failing to interest the manufacturers in UK and US, he licensed his design to Apex Ltd, a Japanese company. His design became hugely popular in Japan. With this success, he started manufacturing his machine in the UK. Upon its release there, it became hugely popular and was subsequently introduced world over. His determination in persevering through 5127 prototypes has resulted in a billion dollar company. He has continued to introduce improved products like hand dryers, desk fans and hair dryers.

STEP 48

WRITING YOUR PROJECT REPORT

Reading maketh a full man, conference a ready man, and writing an exact man. - Francis Bacon. Essays, chapter '"Of Studies" (1625).

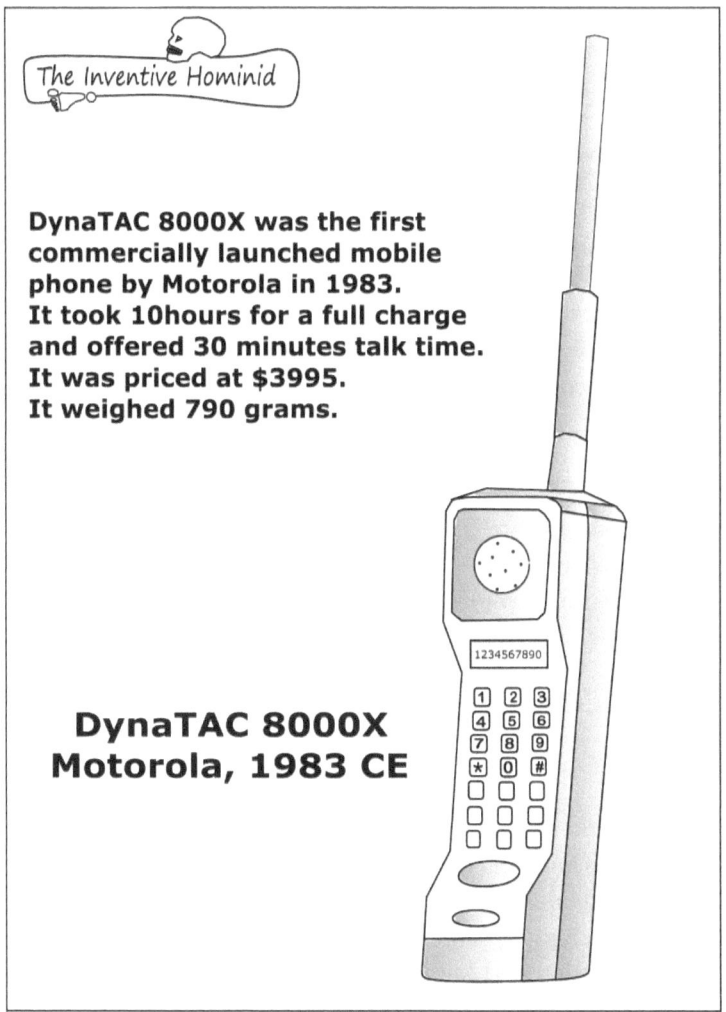

The Inventive Hominid

DynaTAC 8000X was the first commercially launched mobile phone by Motorola in 1983.
It took 10hours for a full charge and offered 30 minutes talk time.
It was priced at $3995.
It weighed 790 grams.

DynaTAC 8000X
Motorola, 1983 CE

TIH - First Mobile Phone (48 IR 1, 48 TR 1)

Learning Objectives

After studying the contents of this chapter, you should be able to:
- Learn about the elements of a project report and a business plan
- Understand the process of writing an effective project report

HOW TO WRITE YOUR PROJECT REPORT?

Decide the purpose of the report; if it is for an academic requirement, then there will be rules and guidelines that you have to adhere.

COMPONENTS OF AN ACADEMIC PROJECT REPORT

A typical academic project report will have the following components:
- ✓ Introduction
- ✓ Project Objectives
- ✓ Literature Review
- ✓ Methodology
- ✓ Experiments Conducted
- ✓ Design
- ✓ Results
- ✓ Analysis
- ✓ Interpretation
- ✓ Conclusion

If you are preparing a report for potential investors, then you need to prepare a business plan

COMPONENTS OF A BUSINESS PLAN

A typical business plan will have the following components:
- ✓ Executive Summary.
- ✓ Description of business/ product.
- ✓ Products and Services to be offered
- ✓ Market analysis/ survey
- ✓ Business Strategy and Implementation plan
- ✓ Organization and Management Team
- ✓ Details of expenses and revenue
- ✓ Funding requirement and plan
- ✓ Sales and revenue projection

Once you have decided the type of report you want to write, the next step is to start writing your report. Just dive in and start typing, there is no magic formula to write a perfect report in the first attempt.

Have a plan of what you want to write in each of the components. Then just start typing as words form in your mind, just fill up all the chapters/ components.

Once you have filled up all the sections, then start reading from the beginning and keep adding your thoughts to strengthen areas and cut out sections, which do not gel.

After a number of revisions, the report will start to take a wonderful shape.

In a group project, make sure everybody's ideas are considered, and leave it to the most imaginative person to type the report.

Alternatively, Assign / divide various components of the report among group members and let everybody use their imagination to enrich the chapters.

Make sure you use the graphical tools effectively to create graphs, figures, flowcharts, etc. to make the report attractive and readable.

Once your draft report is ready, get opinions from your project advisor/ mentor or well-wishers. Remember, ultimately a third person who reads your report should get the understanding of what you are trying to convey.

Most reports are not read beyond the executive summary. Therefore, the executive summary should be to the point, and hook the reader into going beyond it.

AVOIDING PLAGIARISM

What is plagiarism? If you use others ideas, words, figures, etc. without acknowledging them, then it constitutes plagiarism. (Please refer chapter 45)

USING AUTHENTIC SOURCE OF INFORMATION

When you bolster your statements with references, you should be careful to use only reliable sources of information.

Do not just cite information from dubious websites, books, and journals.

Cite and use information only from reputed/ peer reviewed sources of information respected and accepted in your field.

Do not directly cite Wikipedia, but try to retrieve the original source of information normally listed at the end of each Wikipedia article, verify, and cite them.

ACKNOWLEDGING YOUR COLLABORATORS

When you work on a group task like a project, it is important that you should not hog the entire credit for yourself.

List all the collaborators who helped to achieve the task.

Acknowledge all the people who helped you, no matter how little the contribution might be, right from the idea stage to submitting your report in the acknowledgment section of your report.

Inspiring Inventors

The man who lived and built the Gandhian Way

Lawrence Wilfred "Laurie" Baker (1917-2007) was a British born Indian Architect renowned for incorporating Gandhian Principles in Architecture. Laurie baker was born in England in a devout Christian family and later became a Quaker. After graduation, he chose to serve in the Friends Ambulance Unit during the Second World War. While serving the medical team treating casualties in the China-Japan war, his health took a toll and decided to return to England to recuperate. On his way to England, he waited in Bombay for a few days waiting for his ship [48.1].

His life took a turn, when he met Gandhiji and was profoundly impressed by his philosophy of non-violence and simple lifestyle. He decided to stay back in India and contributed to the Leprosy Mission, building treatment centres. There he met an Indian doctor Elizabeth Jacob and finding much in common, they got married in 1948. When they travelled to Pithorgarh after marriage, they found that the people there badly needed medical assistance. They built a hospital there and stayed for 16 years, before returning to Kerala [48.2].

Deeply impacted by the profound sustainable principle of Gandhiji's philosophy, that people in India should build a house with materials found within a radius of 5 kilometers, he set out practicing it in his buildings. He started by learning about vernacular architecture from the skilled masons and often would work alongside them in constructing a building. He built buildings, which used only locally available materials such as brick, stone, lime, etc. he avoided unnecessary elements such as plastering and painting the walls, costly windows and doors and used them only when absolutely necessary.

He never altered the topography of the site and used them in building with minimum injury to nature. He embodied principles of passive energy building, long before they became fashionable [48.3]. His signature style such as rattrap bond, exquisite brick jails and blending with nature, elevates architecture closer to nature. Moreover, he lived they way he preached, simple and austere. In fact, his shoes made of waste rags, which he brought from China, fascinated Gandhiji. He was recognized through various awards nationally and internationally. However, a true tribute to him would be to adopt his sustainable architectural principles and spread them everywhere.

STEP 49

PRESENTING YOUR PROJECT ON THE BIG DAY

Speak with such quality that it binds the friends who listen, and casts a spell on even the foes who do not – Thiruvalluvar, Thirukkural- 643.

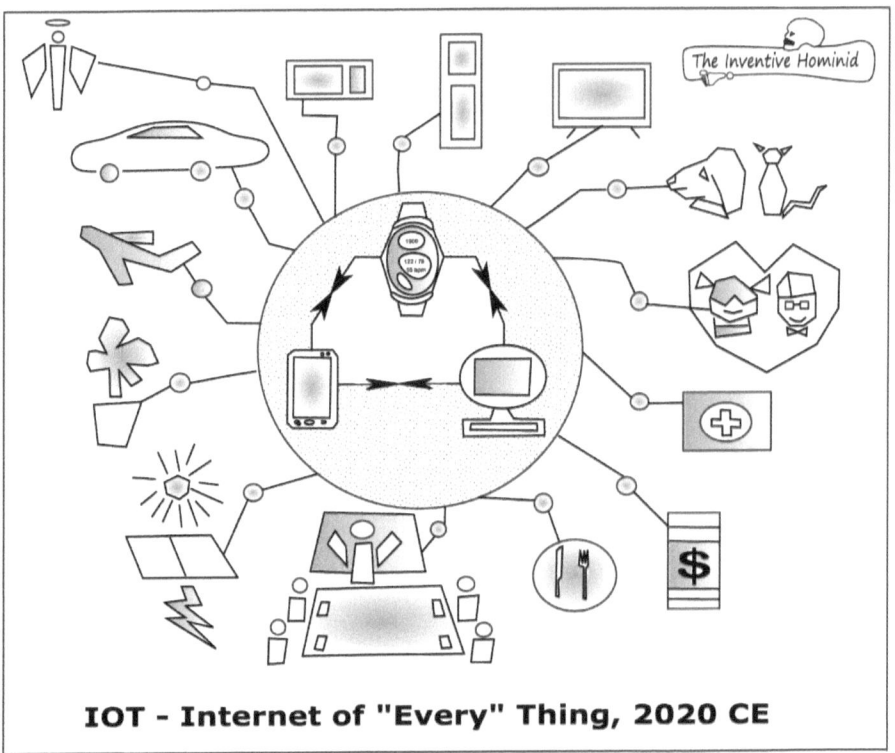

IOT - Internet of "Every" Thing, 2020 CE

TIH - IOT (49 TR 1)

Learning Objectives

After studying the contents of this chapter, you should be able to:
- Learn about the aspects of making an effective project presentation
- Know about tools available to enhance your project presentation
- Understand the attributes employers look for in a graduate

PRESENTING YOUR PROJECT

DEFENDING IT IN ACADEMIA AND PITCHING IT TO EMPLOYERS

A good presentation goes a long way in communicating your project effectively. With the advance of technology, now it is possible with the correct use of multimedia to give a complete picture of your project to the evaluators and the audience.

EVALUATE THE AUDIENCE

If it is an academic requirement exam, then a panel of professors will evaluate it. Professors normally do not tolerate elaborate multimedia enhancements. They would demand a concise explanation of the main achievements in technical terms. So prepare your slides with minimum decoration. Make sure to add the links to references to support your work. Do not forget to brush up your fundamentals in the areas in which your project scope falls.

AFTER THE PROJECT EXAM: PITCHING IT TO EMPLOYERS

Some schools invite companies to their final project presentations. If you are going to present to potential employers, make sure you do research on the background of the companies that are going to be present.

Identify companies of your interest and elaborate on the aspects of your project related to these companies. Make sure you highlight the entire process of technical investigation to show the efforts you have undertaken.

If you are presenting as a group, plan and rehearse the specific portions allotted to each member as per their contribution. Ensure that no group member is unfairly treated and all are given a fair share of the time to present.

PRESENTATION TOOLS WHICH YOU CAN USE 49.1, 49.2

There are a number of websites that provide presentation tools. Some of these are:

- o Prezi
- o Google presentation
- o Glogster Edu
- o Haiku deck
- o helloslide

Search Alert

- o Animoto
- o Kizoa
- o Vuvox
- o Voki
- o Tellagami
- o Slideful
- o Powtoon
- o knovio

There are still a number of other tools freely available, kindly search in the internet and adopt tools which are best suited to the style of your presentation.

WHAT EMPLOYERS WANT FROM POTENTIAL EMPLOYEES?

Yes, the world is getting inhumane day by day and employees are increasingly being considered as head counts and billable resources than human beings. Still majority of us need to get a job to ensure our survival, so what do employers look for in an employee:

- Enthusiasm for the position offered – can be assessed based on your energy levels during the interview
- Basic knowledge, skills, & communication: fundamental knowledge in the technical area of employment, special skills like software certification, etc. above average communication skills
- Cultural fit: whether you are a team player without some weird aspects in your CV

If you tick all these areas, then you will be appointed.

Inspiring Inventors

The mistake that saved many lives and is still saving many more

Wilson Greatbatch (1919-2011) was an American engineer and inventor who held more than 325 patents and was a member of National Inventors hall of Fame. Greatbatch born in buffalo, after completing his schooling, served in the World War II as an aviation chief radioman until 1945. After his military career, at the age of 26, attended Cornell University graduating in Electrical Engineering in 1950 and earned his Master's degree from University of Buffalo in 1957[49.3].

While he was building a device to record heart rhythms, he inadvertently inserted an unsuitable resistor. The device instead of producing oscillating impulses, produced intermittent electrical impulses. Realizing that these intermittent impulses imitated the heart rhythm, he imagined a small device that can stimulate the heart with electrical impulses. Within 2 years, he made a pacemaker of 2 cubic inch in size and tested it on a dog with success. It was implanted into humans soon, but lasted only 3 hours to 2 days.

He left his job to dedicate all his time to develop the pacemaker and using his $2000 savings, collaborated with William Chadack. By 1960, he had a device that worked for 18 months and after using the newly developed Lithium-Iodine battery, it lasted for 10 years or more. Today, the company he started as Greatbatch Inc. is now the most successful producer of pacemakers in the US.

STEP 50

KEEPING THE FIRE BURNING - LIVING AS A GENIUS INVENTOR

What one fool can do, another can. (Ancient Simian Proverb.)

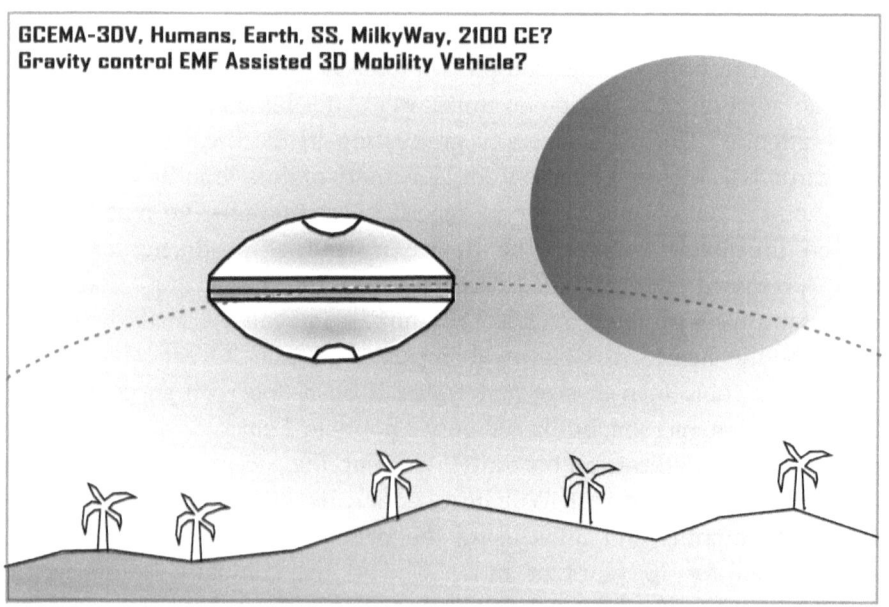

TIH - 3D mobility vehicle – 2100 CE?

Learning Objectives

After studying the contents of this chapter, you should be able to:
- Learn about the requirements of living an inventor's life

KEEPING THE FIRE BURNING

First rule to keep the inventing spirit burning is to adopt solving problems as your life's mission.

You should be ready to grapple with a problem without giving up until the problem cracks open and gives out a solution. We do not appreciate the humongous effort put by Edison in trying out 6000 materials to arrive at a suitable material for the lamp's filament. Most of us would give up after trying 2-3 or 8-10, as we live in the age of instant gratification.

We have to understand Edison's remark "Genius is one percent inspiration and 99 percent perspiration." Inventors spend long hours, days, months, years all alone focused on one single problem. Many have come up with wonderful ideas, but only few have transformed them into useful marketable products. There lies the entrepreneurial spirit of inventors like Edison.

There are people who have successfully invented things while holding a full time day job. They are able to switch off completely, once they finish their daily day job and begin focusing their attention on inventions they work on. Abraham Lincoln with all his activities in different spheres of life is the only President of US to hold a patent. He designed a device for lifting boats over shoals and other obstacles. There have been many examples of people inventing groundbreaking devices while quietly holding on to their professions.

Hence leading the life of inventor needs persistence in trying to find a solution patiently over as long a time it requires. So go ahead, choose the most critical problem faced by humanity and resolve to find a solution even if it takes your entire life.

Happy Inventing!

Moreover, thanks in advance for helping humanity or maybe the entire planet!

𝓘𝓷𝓼𝓹𝓲𝓻𝓲𝓷𝓰 𝓘𝓷𝓭𝓲𝓿𝓲𝓭𝓾𝓪𝓵𝓼

The bridge engineering man whose father wished he never gets a job 50.1

Girish Bharadwaj (born 1950) born in Sullia is known for building around 127 low cost bridges in remote villages across India. In 1973, he graduated in Mechanical Engineering and while he was looking for a job, his father blessed him saying, "May you not succeed in any of the interviews, when you go with your begging bowl (engineering degree)". After attempts at starting an engineering company with partners, he started his own company Ayas Shilpa (meaning Metal Sculpture) in 1975. Until date, he has constructed around 127 suspension bridges, which cost only $1/10^{th}$ the cost of conventional bridges and are constructed in around 3 months as against 3 years for standard concrete bridges. Awarded the Padma Shri in 2017, the bridge man says, "the credit should go to my entire team, and the affection of people is an added bonus."

References and Further Reading

The web links given along with the references may have moved with time and might not work. Hence, readers are advised to copy the entire title of the article, paste it in a search engine, and search to access the article in its latest location.

STEP -1
-1.1   Canons of Ethics for engineers, Engineers' Council for Professional Development. (1947). Canons of ethics for engineers
-1.2   Project Management Institute, What is Project Management? https://www.pmi.org/about/learn-about-pmi/what-is-project-management
-1.3   ABET-EAC, Criteria for Accrediting Engineering Programs, 2019 – 2020, https://www.abet.org/accreditation/accreditation-criteria/criteria-for-accrediting-engineering-programs-2019-2020/#3
-1.4   Engineering, https://www.britannica.com/technology/engineering
STEP 1
1.1   Codex Arundel on the British Library's Catalogue of Illuminated Manuscripts
1.2   David Alan Brown, Leonardo (Da Vinci), Leonardo Da Vinci: Origins of a Genius, Yale University Press, 1998, p. 7, ISBN 0300072465
1 TR 1: Spear: Monte Morin, Stone-tipped spear may have much earlier origin, November 16, 2012.
        http://articles.latimes.com/2012/nov/16/science/la-sci-hafting-spears-20121116
STEP 2
2.1   CAE, Evolution of Engineering Education in Canada, Canadian Academy of Engineering, 1999.
2.2   Mary Bellis, 2017, Samuel Morse and the Invention of the Telegraph, https://www.thoughtco.com/communication-revolution-telegraph-1991939.
2 TR 1 : Boomerang: "Kimberley rock art could be among oldest in the world". ABC News AU.
2 TR 2 : Boomerang: Dan Drollette, Science of Boomerangs: How to Make & Throw the Aussie Magic, Oct 1, 2009, https://www.popularmechanics.com/science/a4809/4219929/
STEP 3

3.1   Kalam, A. P. (2013). My Journey, Transforming Dreams in to Action. New Delhi: Rupa Publications India Pvt.LTD

3.2   Kalam, A. P. (2002). Ignited Minds. Gurgaon: Penguin Books India Pvt. Ltd.

3.3   My profile, http://www.abdulkalam.com/kalam/theme/jsp/guest/myprofile.jsp

3 IR 1 : Harpoon adapted from: Otto Ludvig Sinding (1842-1909), Norwegian painter etc (Popular Science Monthly Volume 46) [Public domain], via Wikimedia Commons https://commons.wikimedia.org/wiki/File:PSM_V46_D468_A_kayak_man_attacked_by_a_walrus.jpg

3 TR 1 : Harpoon: Forbes, Jack D. (1998). In The American Discovery of Europe. University of Illinois Press. p. 103. ISBN 0-252-03152-0. Google Book Search.

STEP 4

4.1   What are the benefits of group work? Eberly Center-Teaching Excellence & Educational Innovation, Carnegie Mellon University, https://www.cmu.edu/teaching/designteach/design/instructionalstrategies/groupprojects/benefits.html

4.2   Chitra Reddy, Top 16 Advantages and Disadvantages of Working in a Group, https://content.wisestep.com/top-advantages-and-disadvantages-of-working-in-a-group/

4.3   "Joint Intelligence Objectives Agency". U.S. National Archives and Records Administration. https://www.archives.gov/iwg/declassified-records/rg-330-defense-secretary

4.4   "Operation "Osoaviakhim"". Russian space historian Anatoly Zak. http://www.russianspaceweb.com/a4_team_moscow.html

4 TR 1 : Indus : Singh, Upinder (2008). A History of Ancient and Early medieval India : from the Stone Age to the 12th century. New Delhi: Pearson Education. p. 137. ISBN 978-8131711200.

4 TR 2 : Indus: Chakrabarti, D.K. (2004). Indus Civilization Sites in India: New Discoveries. Mumbai: Marg Publications. ISBN 978-8185026633.

4 IR 1 : adapted from: Indus Seals, By MrABlair23 - Own work, Public Domain, https://commons.wikimedia.org/w/index.php?curid=11058028

4 IR 2 : adapted from: Pre-History & Archaeology, Indus Valley Civilization, http://nationalmuseumindia.gov.in/prodCollections.asp?pid=44&id=1&lk=dp1

STEP 5

5.1   Walker, John (1781?-1859)". Dictionary of National Biography. London: Smith, Elder & Co. 1885–1900.

5.2   John Walker's Friction Light, http://www.bbc.co.uk/ahistoryoftheworld/objects/hQR9oN5LTeCLcuKfPDMJ9A

5.3   William H. Doolittle (1903). Inventions in the century. London: W. & R. Chambers.

5.4   George Iles (1912). "Cyrus H. McCormick". Leading American Inventors (2nd ed.). New York: Henry Holt and Company. pp. 276–314.

STEP 6

6.1   McGann, James G., "2017 Global Go To Think Tank Index Report" (2018).TTCSP Global Go To Think Tank Index Reports. 13. https://repository.upenn.edu/think_tanks/13

6.2   The 100 Most Influential Think Tanks In The World For 2017, Dr. Amarendra Bhushan Dhiraj, January 31, 2017

6.3   Homans, James E., ed. (1918). "Edison, Thomas Alva". The Cyclopædia of American Biography. New York: The Press Association Compilers, Inc.

6.4   Thomas Edison, in Frank Dyer and Lewis Martin, Edison: His Life and Inventions, xxiv. Available at: gutenberg.org/eBooks/820.

6 TR 1 : Pyramid : Levy, Janey (2005). The Great Pyramid of Giza: Measuring Length, Area, Volume, and Angles. Rosen Publishing Group.ISBN 1404260595.

6 IR 1 : adapted from: Pharaoh, By Jeff Dahl - Own work, CC BY-SA 4.0, https://commons.wikimedia.org/w/index.php?curid=3307561 https://en.wikipedia.org/wiki/User:Jeff_Dahl?rdfrom=commons:User:Jeff_Dahl#/media/File:Pharaoh.svg

STEP 7

7.1   UN Sustainability goals, http://www.un.org/sustainabledevelopment/sustainable-development-goals/

7.2   Global Warming of 1.5 °C, an IPCC special report, The Intergovernmental Panel on Climate Change, https://www.ipcc.ch/index.htm

7.3   Global Footprint Network, https://www.footprintnetwork.org/

7 TR 1 : ball : Ortíz C., Ponciano; Rodríguez, María del Carmen (1999) "Olmec Ritual Behavior at El Manatí: A Sacred Space" in Social Patterns in Pre-Classic Mesoamerica, eds. Grove, D. C.; Joyce, R. A., Dumbarton Oaks Research Library and Collection, Washington, D.C., pp. 225–254

7 TR 2 : ball: Cartwright, Mark. "The Ball Game of Mesoamerica." Ancient History Encyclopedia. Last modified September 16, 2013. https://www.ancient.eu/article/604/.

7 IR 1 : ball adapted from Luis Miguel Bugallo Sánchez https://commons.wikimedia.org/wiki/File:2010._Chich%C3%A9n_Itz%C3%A1._Quintana_Roo._M%C3%A9xico.-8.jpg

7 IR 2 : ball adapted from: Michael Auld, SUNDAY, JUNE 22, 2014, Football: Is it American? What role did the Olmec play in this international game? http://yamaye1.blogspot.com/2014/06/football-is-it-american-what-role-did.html

STEP 8

8.1   The Project Gutenberg EBook of 'Gutenberg, and the Art of Printing', by Emily Clemens Pearson

8.2   McLuhan, Marshall (1962). The Gutenberg Galaxy: The Making of Typographic Man (1st ed.). University of Toronto Press. ISBN 978-0-8020-6041-9.

8.3   "How Gutenberg Changed the World". livescience.com.

8 IR 1 : adapted from : Phoenician Ships, Entombet, January 13, 2014 in Age of Sail Warships, https://forum.worldofwarships.eu/topic/2512-phoenician-ships/

STEP 9

9.1   Women in Chemistry – Stephanie Kwolek". Science History Institute.

9.2   Stephanie Kwolek (b. 1923) , American Chemical Society, https://www.acs.org/content/acs/en/education/whatischemistry/women-scientists/stephanie-kwolek.html

9 IR 1 : Kodumanal adapted from: Kodumanal Archeological Site https://tamilnadu-favtourism.blogspot.com/2015/09/kodumanal-archeological-site.html

9 TR 1 : Srinivasan, Sharada (15 November 1994). "Wootz crucible steel: a newly discovered production site in South India". Papers from the Institute of Archaeology. 5: 49–59. doi:10.5334/pia.60.

9 TR 2, IR 1: Sasisekharan, B. (1999). "Technology of Iron and Steel in Kodumanal"(PDF). Indian Journal of History of Science. 34 (4).

9 IR 2 : Kodumanal adapted from : By Rich Bowen from Lexington, KY, USA (Damascus steel hunting knife) [CC BY 2.0 (https://creativecommons.org/licenses/by/2.0)], via Wikimedia Commons https://commons.wikimedia.org/wiki/File:Damascus_steel_hunting_knife_(4122685428).jpg

STEP 10

10.1   The Project Gutenberg EBook of The Art of Preserving All Kinds of Animal and Vegetable Substances for Several Year, by M. Appert, 1812, http://www.gutenberg.org/files/52551/52551-h/52551-h.htm

10.2   Garcia, Rebeca; Adrian, Jean (2009), "Nicolas Appert: Inventor and Manufacturer" (https://www.researchgate.net/publication/240546659_Nicolas_Appert_Inventor_and_Manufacturer), Food Reviews International, 25, (2), retrieved 2017-10-11

10 IR 1 : lighthouse adapted from: Prof. Hermann Thiersch (1874–1939) (Hermann Thiersch) [Public domain], via Wikimedia Commons, https://commons.wikimedia.org/wiki/File:Lighthouse_-_Thiersch.png

10 TR 1 : McKenzie, Judith (2011). The Architecture of Alexandria and Egypt: 300 BC – AD 700. Yale University Press. p. 42. ISBN 978-0300170948.

STEP 11

11.1   "Venkatraman Ramakrishnan - Biography: From Chidambaram to Cambridge: A Life in Science". Stockholm: nobelprize.org.

11.2   The Royal Society, https://royalsociety.org/people/venki-ramakrishnan-12139/

11 IR 1 : crossbow adapted from: Yprpyqp [CC BY-SA 4.0 (https://creativecommons.org/licenses/by-sa/4.0)], from Wikimedia Commons, https://commons.wikimedia.org/wiki/File:Qin_crossbow.jpg

11 TR 1 : Needham, Joseph (2004), Science and Civilization in China, Vol 5 Part 6, Cambridge University Press, p. 135, ISBN 0-521-08732-5

11 TR 2 : Gallwey, Sir Ralph (1990). "The Crossbow" (Ninth Impression ed.). The Holland Press. p. 337.

11 TR 3 : Lin, Yun. "History of the Crossbow," in Chinese Classics & Culture, 1993, No.4: p. 33–37.

STEP 12

12.1   Ffrench, Yvonne. The Great Exhibition; 1851. London: Harvill Press, 1950.

12.2   The Project Gutenberg eBook, The World's Fair, by Anonymous

12 IR 1 : Water wheel adapted from : By W. N. Manning, Photographer [Public domain], via Wikimedia Commons, https://commons.wikimedia.org/wiki/File:Mill_with_Water_Wheel,_Aderholdt%27s_Mill_Road,_Anniston_vicinity_(Calhoun_County,_Alabama).jpg

12 TR 1 : Terry S, Reynolds, Stronger than a Hundred Men; A History of the Vertical Water Wheel. Baltimore; Johns Hopkins University Press, 1983.

STEP 13

13.1   "IPTO – Information Processing Techniques Office", The Living Internet, Bill Stewart (ed), January 2000. https://www.livinginternet.com/i/ii_ipto.htm

13 IR 1: Sluice adapted from: Uditha Wijesena, BISO KOTUWA – A Sri Lankan Engineering Marvel , Wednesday, May 7, 2014 https://udithawijesena.blogspot.com/2014/05/biso-kotuwa-sri-lankan-engineering.html

13 TR 1 : A.Y.Tharshikan , S.G.L.M. Fernando and H.P.V.Vithana, Flow Behaviour within Bisokotuwa Using Physical Model and CFD Model, 5th International Symposium on Advances in Civil and Environmental Engineering Practices for Sustainable Development (ACEPS-2017), http://www.dcee.ruh.ac.lk/images/donaimage/ACEPPProceeding2017/River_Engineering_and_Hydroinformatis/Flow_Behaviour_within_Bisokotuwa.pdf

STEP 14

14.1   Vincent, Julian F. V.; et al. (22 August 2006). "Biomimetics: its practice and theory". Journal of the Royal Society Interface. 3 (9): 471–482.

14.2   "Joy of Manufacturing". Honda History. Honda Motor Co., Ltd.

14.3   Tremayne, David (17 January 2001). "Soichiro Honda: The man behind a legend". Grandprix.com. Inside F1, Inc.

14 IR 1: adapted from steam turbine, By The entry under Aeolipile in volume one of this work states "The cut is copied from Hero's "Spiritalia", edited by Woodcroft, of London." [Public domain], via Wikimedia Commons, https://commons.wikimedia.org/wiki/File:Aeolipile_illustration.png

14 TR 1 : Hero (1851) [reprint of 1st century CE original], "Section 50 – The Steam Engine", written at Alexandria, Pneumatica, translated by Bennet Woodcroft, London: Taylor Walton and Maberly, http://himedo.net/TheHopkinThomasProject/TimeLine/Wales/Steam/URochesterCollection/Hero/section50.html

STEP 15

15.1   Deloitte Tech Trends 2018, 2017, 2016, Tech Trends Archive, https://www2.deloitte.com/insights/us/en/focus/tech-trends.html

15.2   2018 Emerging Tech Trends Report, Future Today Institute, https://futuretodayinstitute.com/futures-research/

15.3   UNCTAD, Technology and innovation report 2018: harnessing frontier technologies for sustainable development, 15/05/2018, https://unctad.org/en/PublicationsLibrary/tir2018_en.pdf

15.4   Friedrich Gottlob Keller, 2010 Paper Industry International Hall of Fame Inductee, https://www.paperdiscoverycenter.org/inductees/friedrich-gottlob-keller/

15.5   Burger, Peter (http://www.charlesfenerty.ca/book.html).
Charles Fenerty and his Paper Invention.Toronto: Peter Burger, 2007. ISBN 978-0-9783318-1-8.

15 IR 1 : Catapult adapted from: By Rpanjwani3 [CC BY-SA 3.0 (https://creativecommons.org/licenses/by-sa/3.0)], from Wikimedia Commons, https://commons.wikimedia.org/wiki/File:Mang2.png

15 TR 1 : Purton, Peter (2009), A History of the Early Medieval Siege c.450-1200, The Boydell Press

STEP 16

16.1   "One Hundred Year Study on Artificial Intelligence (AI100)," Stanford University, accessed August 1, 2016, https://ai100.stanford.edu

16.2   NSTC, Preparing for the Future of Artificial Intelligence", prepared by the National Science and Technology Council (NSTC), 2006.

16.3   Making the AI revolution work for everyone. The Future Society, AI Initiative. Nicolas MIAILHE., Cyrus HODES, 2017. http://ai-initiative.org/

16.4   Pettinger, Tejvan. "Biography of Alan Turing", Oxford, UK.

16.5   "Alan Turing: The codebreaker who saved 'millions of lives'". BBC News Technology.

16 TR 1 : James A. O'Kon (2005). "Computer Modeling of the Seventh Century Maya Suspension Bridge at Yaxchilan". Computing in Civil Engineering, Proceedings of the 2005 ASCE International Conference on Computing in Civil Engineering Cancun, Mexico. 179: 124. doi:10.1061/40794(179)124.

16 IR 1 :Suspension bridge adapted from: Resource: Just how advanced Were the Maya? Jim O'Kon's reconstructions of the lost suspension bridge , http://www.mexicolore.co.uk/maya/teachers/resource-just-how-advanced-were-they

STEP 17

17.1   World Economic Forum, MM Buehler, Shaping the Future of Construction: A Breakthrough in Mindset and Technology, 2016, World Economic Forum

17.2   WEF, Shaping the Future of Construction- Future Scenarios and Implications for the Industry, 2018, http://www3.weforum.org/

17.3   A J Francis, The Cement Industry 1796-1914: a History, David & Charles, 1977, ISBN 0-7153-7386-2

17.4   Autobiography of Isaac Charles Johnson Esq, JP, Farncombe & Sons, London, 1912.

17 IR 1: Zero adapted from: J.E. Montucla [Public domain], via Wikimedia Commons, https://commons.wikimedia.org/wiki/File:EuropeanFormOfArabianDigits.png

17 TR 1 : Jan Gullberg (1997), Mathematics: From the Birth of Numbers, W.W. Norton & Co., ISBN 978-0393040029,

STEP 18

18.1   WEF, These are the top 10 emerging technologies of 2016, 2017, 2018. https://www.weforum.org/agenda/archive/emerging-technologies/

18.2   Policy Horizons Canada, 2014, MetaScan 3: Emerging Technologies, http://www.horizons.gc.ca/en/content/metascan-3-emerging-technologies-0

18.2   Founder of UCSD Bioengineering Program. jacobsschool.ucsd.edu. 1 Mar 2004.

18.3   "Bioengineering". Encyclopedia Britannica.

18.4   2050_Global_Agriculture, http://www.fao.org/fileadmin/templates/wsfs/docs/Issues_papers/HLEF2050_Global_Agriculture.pdf

18.5   Bawa, A.S.; Anilakumar, K.R. (2012-12-19). "Genetically modified foods: safety, risks and public concerns—a review". Journal of Food Science and Technology. 50 (6): 1035–1046.

18.6   Genome and genetics timeline – 1973. Genome news network. www.genomenewsnetwork.org/resources/timeline/

18 IR 1: wheelbarrow adapted from: Zhang Zeduan (1085–1145) - http://depts.washington.edu/chinaciv/painting/4ptgqmsh.htm, Fu Xinian, ed. Zhongguo meishu quanji, Liang Song huihua, shang (Series Vol. 3), pl. 51, pp. 128-137 (This applies to all five sections). Collection of the National Palace Museum, Beijing., Public Domain, https://commons.wikimedia.org/w/index.php?curid=3507861

18 TR 1: Needham, Joseph (1965). Science and Civilization in China: Volume 4, Physics and Physical Technology, Part 2, Mechanical Engineering; rpr. Taipei: Caves Books Ltd.

18 IR 2 : wheel barrow adapted from: Isabella Lucy Bird - Chinese Picture: notes on photographs made in China, Public Domain, https://commons.wikimedia.org/w/index.php?curid=30959543

STEP 19

19.1   James L. ColemanJr. The American whale oil industry: A look back to the future of the American petroleum industry? , Natural Resource Research

(1995) 4: 273. https://doi.org/10.1007/BF02257579, September 1995, Volume 4, Issue 3, pp 273–288

19 IR 1 : Moche Gold adapted from: By PIERRE ANDRE LECLERCQ - Own work, CC BY-SA 4.0, https://commons.wikimedia.org/w/index.php?curid=46486071

19 TR 1 : New Perspectives on Moche Metallurgy: Techniques of Gilding Copper at Loma Negra, Northern Peru, Heather Lechtman, Antonieta Erlij and Edward J. Barry, Jr., American Antiquity, Vol. 47, No. 1 (Jan., 1982), pp. 3-30, Published by: Cambridge University Press, DOI: 10.2307/280051, https://www.jstor.org/stable/280051

STEP 20

20.1   Arthur D Little, The Future of Mobility 3.0, Reinventing mobility in the era of disruption and creativity, 2018. http://www.adlittle.com/futuremobilitylab/

20.2   Deloitte, The future of mobility- How transportation technology and social trends are creating a new business ecosystem, Deloitte University press. www2.deloitte.com/

20.3   The Project Gutenberg EBook of My Life and Work, by Henry Ford

20.4   Ford R. Bryan, "The Birth of Ford Motor Company" Archived August 29, 2012, at the Wayback Machine., Henry Ford Heritage Association, retrieved August 20, 2012.

20.5   The Showroom of Automotive History: 1896 Quadricycle Archived June 15, 2010, at the Wayback Machine.

20 IR 1 :Charkha adapted from: By Jigar Brahmbhatt [CC BY-SA 3.0 (https://creativecommons.org/licenses/by-sa/3.0)], from Wikimedia Commons https://commons.wikimedia.org/wiki/File:Charkha_kept_at_Gandhi_Ashram.jpg

20 TR 1: Smith, C. Wayne; Cothren, J. Tom (1999). Cotton: Origin, History, Technology, and Production. 4. John Wiley & Sons. pp. viii. ISBN 978-0471180456.

STEP 21

21.1   Life Extension, Wikiversity, en.wikiversity.org/wiki/Life_extension

21.2   Rick Hammer, 17 Amazing Healthcare Technology Advances of 2017, 2017, Https://Getreferralmd.Com/2017/01/17-Future-Healthcare-Technology-Advances-Of-2017-Referralmd/

21.3   OLDLIST, a database of old trees, http://www.rmtrr.org/oldlist.htm

21.4   Rozell (2014) "Bowhead Whales May Be the World's Oldest Mammals", Alaska Science Forum, https://alaskatrekker.com/2014/12/bowhead-whales/

21.5   BBC News – South Asia (2006-03-23). "'Clive of India's' tortoise dies." BBC News. BBC Online. http://news.bbc.co.uk/2/hi/south_asia/4837988.stm.

21 IR 1: Hagia Sophia adapted from: Steve Evans from Citizen of the World (Istanbul 036) [CC BY 2.0 (https://creativecommons.org/licenses/by/2.0)], via Wikimedia Commons
https://commons.wikimedia.org/wiki/File:Istanbul_036_(6498284165).jpg

21 IR 2: Hagia Sophia adapted from: See page for author [Public domain], via Wikimedia Commons
Wilhelm Lübke / Max Semrau: Grundriß der Kunstgeschichte. 14. Auflage. Paul Neff Verlag, Esslingen, 1908; German Wikipedia, original upload 28. Aug 2004 by Rainer Zenz
https://commons.wikimedia.org/wiki/File:Hagia-Sophia-Laengsschnitt.jpg

21 TR 1: Fazio, Michael; Moffett, Marian; Wodehouse, Lawrence (2009). Buildings Across Time (3rd ed.). McGraw-Hill Higher Education. ISBN 978-0-07-305304-2.

STEP 22

22.1   Misra, Ajay K, Emerging Materials Technologies That Matter to Manufacturers, NASA Glenn Research Center, Cleveland, OH United States, Conference Paper Manufacturing Matters 2015; 26 Feb. 2015; Milwaukee.
https://ntrs.nasa.gov/archive/nasa/casi.ntrs.nasa.gov/20150022404.pdf

22.2   WEF, These are the top 10 emerging technologies of 2016, 2017, 2018.
https://www.weforum.org/agenda/archive/emerging-technologies/

22.3   Deloitte Tech Trends 2018, 2017, 2016, Tech Trends Archive,
https://www2.deloitte.com/insights/us/en/focus/tech-trends.html

22.4   LED developed in Russia in 1920s, By Nick Farrell, The Inquirer,
http://russianpatentsblog.patentsfromru.com/2007/04/12/youd-better-not-ignore-russian-prior-art-search-says-new-scientist-magazine/

22.5   Zheludev, Nikolay (April 2007). "The life and times of the LED – a 100-year history" (PDF). Nature Photonics. Nature Publishing Group. 1 (4): 189–192.

22.6   M. A. Novikov (January 2004) "Oleg Vladimirovich Losev: Pioneer of Semiconductor Electronics," Physics of the Solid State, vol. 46, no. 1, page 1-4

22.7   Nano-Bio-Info-. Cogno. Innovations: Converging. Technologies In. Society edited by. William Sims Bainbridge. National Science Foundation., Springer.

22.8   National Aeronautics and Space Administration, https://www.nasa.gov/

22.9   Chronology of Space Exploration, http://spacechronology.com/

22.10   Indian Space Research Organisation, https://www.isro.gov.in/missions

22 IR 1: Wind mill adapted from: Saupreiß [CC BY-SA 3.0 (https://creativecommons.org/licenses/by-sa/3.0)], from Wikimedia Commons https://commons.wikimedia.org/wiki/File:Persische_Windm%C3%BChle_Model_-_Deutsches_Museum_M%C3%BCnchen.jpg

22 TR 1: Eldridge, Frank (1980). Wind Machines (2nd ed.). New York: Litton Educational Publishing, Inc. p. 15. ISBN 0-442-26134-9.

STEP 23

23.1   Frank Lewis Dyer and Thomas Commerford Martin, The Project Gutenberg EBook of Edison, His Life and Inventions, 2006

23.2   Shulman, Seth (1999). Owning the Future. Houghton Mifflin Company. pp. 158–160.

23 IR 1 : Apartments adapted from: Lorax [CC-BY-SA-3.0 (http://creativecommons.org/licenses/by-sa/3.0/)], via Wikimedia Commons https://commons.wikimedia.org/wiki/File:Mesaverde_cliffpalace_20030914.752.jpg

23 TR 1: Fagan, B. (2005), Chaco Canyon: Archaeologists Explore the Lives of an Ancient Society, Oxford University Press, ISBN 0-19-517043-1

STEP 24

24.1   Stories of Great Inventors Fulton, Whitney, Morse, Cooper, Edison, Project Gutenberg's Stories of Great Inventors, by Hattie E. Macomber, https://www.gutenberg.org/files/19533/19533-h/19533-h.htm

24 IR 1: Nalanda adapted from: Ankitnirala [CC BY-SA 4.0 (https://creativecommons.org/licenses/by-sa/4.0)], from Wikimedia Commons https://commons.wikimedia.org/wiki/File:Nalanda_University_Full_View.jpg

24 TR 1: Scharfe, Hartmut (2002). Education in Ancient India. Handbook of Oriental Studies. 16. Brill. ISBN 9789004125568.

STEP 25

25.1   Project Management Institute, What is Project Management?, https://www.pmi.org/about/learn-about-pmi/what-is-project-management

25.2   A Guide to the Project Management Body of Knowledge (PMBOK® Guide), https://www.pmi.org/pmbok-guide-standards/foundational/pmbok

25.3   Sprinkling water using Chandraprabha rain gun, MJ Prabhu, The Hindu, 10 March 2011, https://www.thehindu.com/todays-paper/tp-features/tp-sci-tech-and-agri/Sprinkling-water-using-Chandraprabha-rain-gun/article14941451.ece

25.4   The National Innovation Foundation (NIF) – India, http://nif.org.in/aboutnif

25.5   The HoneyBee Network, http://www.honeybee.org/

25.6   Society for Research and Initiatives for Sustainable Technologies and Institutions
(SRISTI), http://www.sristi.org/

25.7   Grassroots Innovation and Augmentation network (GIAN), http://www.gian.org/

25 IR 1: Torpedo adapted from: Hasan Al-Rammah's torpedo, from Islamic Technology: An Illustrated History, by al-Hassan, Ahmad Y., and Donald R., Hill.

25 TR 1: Hassan Al Rammah, https://weaponsandwarfare.com/2015/10/18/hassan-al-rammah/
Model, Rocket Torpedo, Hassan er-Rammah or Hasan al-Rammah, ca. 1280 A.D.
https://airandspace.si.edu/collection-objects/model-rocket-torpedo-hassan-er-rammah-or-hasan-al-rammah-ca-1280-ad

STEP 26

26.1   "Padma Awards 2016: Complete list". Time of India. Retrieved August 31, 2016.

26.2   Michael Noer , One Man, One Computer, 10 Million Students: How Khan Academy Is Reinventing Education". Forbes. www.forbes.com

26.3   Dreifus, Claudia (2014). " It All Started With a 12-Year-Old Cousin". The New York Times.
https://www.nytimes.com/2014/01/28/science/salman-khan-turned-family-tutoring-into-khan-academy.html

26 IR 1: Aztec Armour adapted from : Unknown - Foundation for the Advancement of Mesoamerican Studies, Inc.[1] Original document held at the Bodleian Library, Oxford. Shelfmark: MS. Arch. Selden. A. 1., Public Domain, https://commons.wikimedia.org/w/index.php?curid=1429083

26 IR2, TR 2: Cotton Armor vs. Steel, https://pintsofhistory.com/2011/08/10/mesoamerican-cotton-armor-better-than-steel/

26 TR 1: The Mexican (Aztec) version of the bulletproof vest, http://freyathefrypan.blogspot.com/2017/09/the-mexica-aztec-version-of-bulletproof.html

STEP 27

27.1   Elon Musk, American entrepreneur, Written By: Erik Gregersen, Last Updated: Nov 21, 2018, https://www.britannica.com/biography/Elon-Musk

27.2   Billionaire Elon Musk says he was 'raised by books' and credits his success to these 8, Zameena Mejia | Marguerite Ward, 16 Nov 2017, https://www.cnbc.com/2017/11/16/tesla-ceo-elon-musk-says-he-was-raised-by-books.html

27 IR 1: Inca Highway adapted from : Sayacmarka, ruins on the Inca Trail. By Mx._Granger - Own work, CC0, https://commons.wikimedia.org/w/index.php?curid=64983343

27 TR 1: D'Altroy, Terence N. (2002). The Incas. Blackwell Publishers Inc. ISBN 0-631-17677-2

STEP 28

28.1   Raewyn Connell et al. (1975). How to do small surveys – a guide for students in sociology, kindred industries and allied trades. School of Social Sciences. Flinders University. p. 1.

28.2   History of Philosophical Transactions, https://arts.st-andrews.ac.uk/philosophicaltransactions/brief-history-of-phil-trans/

28 IR 1, TR 1: Snowshoes and the Canadian First Nations, Written by Nigel Boney, Published on June 17, 2012, https://www.snowshoemag.com/2012/06/17/snowshoes-and-the-canadian-first-nations/

STEP 29

29.1   Ballpoint Pen: Fascinating Facts about the Invention, http://www.ideafinder.com/history/inventions/ballpen.htm

29.2   M. Frumkin, "The Origin of Patents", Journal of the Patent Office Society, March 1945, Vol. XXVII, No. 3, http://www.compilerpress.ca/Library/Frumkin%20Origin%20of%20Patents%20JPOS%201945.htm

29.3   Article 27.1. of the TRIPs Agreement.

29.4   Incandescent Lamps, History of the Incandescent Light (1802 — today), EdisonTechCenter.org http://edisontechcenter.org/incandescent.html

29 TR 1 : Native American Technology and Art, Willow Branches and Other Twigs and Roots. http://www.nativetech.org/willow/willow.htm

STEP 30

30.1   Values first, An Interview with Narayana Murthy,
        http://www.indiaseminar.com/2000/485/485%20interview.htm
30 TR 1 : Aĭkhenvald, Alexandra (2012). Languages of the Amazon. Oxford
        University Press. p. 64.
STEP 31
31.1   Biography of Dr. V Kurien, http://www.drkurien.com/biography
31.2   Damodaran, Harish (13 September 2004). "Amul's tech wizard, Dalaya
        passes away". The Hindu Business Line.
31.3   Dr. V Kurien, father of white revolution passes away, Anand, Sept 09,2012,
        http://www.amul.com/m/dr-v-kurien
31 IR1, TR 1: Igloo, https://www.thecanadianencyclopedia.ca/en/article/igloo#
STEP 32
32.1   Karanjia, Burjor Khurshedji (2004). Vijitatma: Pioneer-founder Ardeshir
        Godrej. Bombay: Penguin. ISBN 0-670-05762-2.
32.2   Pradeep Puri, A Peep into the Godrej Philosophy of Growth,
        www.business-standard.com,
        https://www.business-standard.com/article/specials/a-peep-into-the-
        godrej-philosophy-of-growth-198021801139_1.html
32 TR 1 : Pascaline calculator, Scientific instruments at the Musée des Arts et
        Métiers
32 TR 2 : Things that Count,
        http://metastudies.net/pmwiki/pmwiki.php?n=Site.Introduction
32 IR 1 : Calculator adapted from : Rama [CC BY-SA 3.0 fr
        (https://creativecommons.org/licenses/by-sa/3.0/fr/deed.en),
        https://commons.wikimedia.org/wiki/File:Pascaline-CnAM_823-1-
        IMG_1506-black.jpg

STEP 33
33.1   How it all Began 1844: Carl Benz, https://www.mercedes-
        benz.com/en/mercedes-benz/classic/history/corporate-history/
33.2   Doug Nye, Carl Benz, and the motor car, Priory Press Ltd, 1973.
33 IR 1: Rocket adapted from: By Home, Robert, born 1752 - died 1834 (artist)
        [Public domain], via Wikimedia Commons
        https://commons.wikimedia.org/wiki/File:Indian_soldier_of_Tipu_Sultan
        %27s_army.jpg
33 TR 1: The Battle of Pollilur 1780: D/ Tipu's archers, The National Galleries of
        Scotland,
        https://web.archive.org/web/20070929195356/http://www.tigerandthistle.n
        et/tipu312.htm
STEP 34

34.1 Project Gutenberg's A Series of Lessons in Raja Yoga, by Yogi Ramacharaka, 1906, The Yogi Publication Society. 2004.

34.2 Eido Michael Luetchford, Introduction to Buddhism and the Practice of Zazen, Windbell Publications, 2000, https://www.holybooks.com/introduction-buddhism-practice-zazen/

34.3 Tao Te Ching by Lao-tzu - Translated by J. Legge, https://www.sacred-texts.com/tao/taote.htm

34.4 Peter Douris et. al., Martial Art Training and Cognitive Performance in Middle-Aged Adults, Journal of Human Kinetics volume 47/2015, 277-283.

34.5 Yogi Ramacharaka, The Hindu-Yogi Science of Breath, A Complete Manual of The Oriental Breathing Philosophy of Physical, Mental, Psychic and Spiritual Development. 2004,

34.6 Rama Prasad, The Science of Breath & the Philosophy of the Tatwas, The Theosophical Publishing Society, London (1890), https://www.holybooks.com/science-breath-philosophy-tatwas/

34.7 Angel, Leslie & Polzella, Donald & C Elvers, Greg. (2010). Background music and cognitive performance. Perceptual and motor skills. 110. 1059-64. 10.2466/pms.110.3c.1059-1064.

34.8 Project Gutenberg's A Practical Guide to Self-Hypnosis, by Melvin Powers, 2007.

34.9 The Project Gutenberg eBook, Your Mind and How to Use It, by William Walker Atkinson, 1911, The Elizabeth Towne Co. 2013.

34.10 Secret habits of the world's best inventors, James Billington, December 4, 20148:48am, 29 Nov 2018, https://www.news.com.au/technology/gadgets/secret-habits-of-the-worlds-best-inventors/news-story/3431735cb28bc7340581239be5f3821d

34.3 A Dave West, Tai Chi Course Book, http://taichibali.com/data/TAI%20CHI%20RETREAT%20COURSE%20%20BOOK%20-%20www.taichibali.com.pdf

34 IR 1: balloon adapted from : Claude-Louis Desrais [Public domain], via Wikimedia Commons https://commons.wikimedia.org/wiki/File:Montgolfier_brothers_flight.jpg

34 TR 1: The Project Gutenberg EBook of Balloons, Airships, and Flying Machines, by Gertrude Bacon. https://www.gutenberg.org/files/54799/54799-h/54799-h.htm

34 TR 2: C.C. Gillispie, The Montgolfier brothers and the invention of aviation 1783-1784, Princeton University Press.

STEP 35

35.1   EU, Training Guidelines - Creative Thinking In Literacy & Language Skills, http://www.creativethinkingproject.eu/training_guide/CTILLS_O1_TrainingGuide_EN.pdf

35.2   Bélanger; Paul; Federighi; Paolo, Unlocking People's Creative Forces: A Transnational Study of Adult Learning Policies, UIE, 2000, ISBN 92-820-1104-6, http://uil.unesco.org/adult-education/

35.3   The Early History of the Airplane, The Wright Brothers' Aeroplane, How We Made the First Flight & Some Aeronautical Experiments. Orville Wright & Wilbur Wright, 2008 [EBook #25420] Project Gutenberg EBook

35 IR 1: locomotive adapted from: circa. 1829 author unknown [Public domain], via Wikimedia Commons, https://commons.wikimedia.org/wiki/File:Trevithick%27s_Coalbrookdale_locomotive,_1803_(British_Railway_Locomotives_1803-1853).jpg

35 TR 1: The Project Gutenberg EBook of A History of the Growth of the Steam-Engine, by
   Robert H. Thurston, http://www.gutenberg.org/files/35916/35916-h/35916-h.htm

STEP 36

36.1   Maday Patent Law, PLLC , How To Document Your Invention, http://www.madaypatentlaw.com/downloads/Documenting%20Your%20Invention.pdf

36.2   Galileo, Italian philosopher, astronomer and mathematician Written By: Albert Van Helden, Last Updated: Oct 26, 2018, https://www.britannica.com/biography/Galileo-Galilei

36 IR 1: Combustion engine adapted from: By Nicéphore Niépce (1765 – 1833) and Claude Niépce (1763 – 1828) [Public domain], via Wikimedia Commons, https://commons.wikimedia.org/wiki/File:Pyreolophore.JPG

36 TR 1 : The life of Nicéphore Niépce, http://www.photo-museum.org/life-nicephore-niepce/

STEP 37

37.1   Open Source Software Directory, The greatest collection of open source and free software, https://opensourcesoftwaredirectory.com/Scientific/Statistical-tools

37.2   Free Open Source Windows Scientific/Engineering Software, https://sourceforge.net/directory/science-engineering/scientific/os:windows/

37.3   Alternatives to Open Source Software Directory for all platforms, https://alternativeto.net/software/open-source-software-directory/

37.4   Rachel Bartee, 2016, 22 Online Research Tools Every College Student should know about, https://www.collegeraptor.com/find-colleges/articles/tips-tools-advice/22-online-research-tools-every-college-student-know/

37.5   Anna Heinrich, 15 Educational Search Engines College Students Should Know About, 03/22/2017, http://www.rasmussen.edu/student-life/blogs/college-life/15-educational-search-engines/

37.6   Antonio Tooley, 10 Great Research Tools for College Students, November 9th, 2015, https://collegepuzzle.stanford.edu/10-great-research-tools-for-college-students/

37.7   Linus Torvalds, https://en.wikipedia.org/wiki/Linus_Torvalds

37.8   Linus Torvalds: A Very Brief and Completely Unauthorized Biography, http://www.linfo.org/linus.html

37.9   Rivlin, Gary. "Leader of the Free World". Wired. Retrieved 30 Nov, 2018., https://www.wired.com/2003/11/linus/

37 IR : Fuel cell adapted from: By Noraneko [Public domain], via Wikimedia Commons https://commons.wikimedia.org/wiki/File:Grove%27s_Gaseous_Voltaic_Battery.png

37 TR 1: "Mr. W. R. Grove on a new Voltaic Combination". The London and Edinburgh Philosophical Magazine and Journal of Science. 1838. https://www.tandfonline.com/doi/abs/10.1080/14786443808649618

STEP 38

38.1   Andy Miah, The A to Z of social media for academia-Your definitive guide to using social media as an academic, March 9, 2017 , https://www.timeshighereducation.com/a-z-social-media

38.2   Ruchira Gupta, "The 100 Most Influential People – Pioneers: Arunachalam Muruganantham". TIME.com. 23 April 2014. Retrieved 26 April 2014., http://time.com/70861/arunachalam-muruganantham-2014-time-100/

38.3   Akila Kannadasan (13 February 2012). "A man in a woman's world". The Hindu. https://www.thehindu.com/features/metroplus/society/a-man-in-a-womans-world/article2875390.ece

38.4   About Jayaashree Industries, http://newinventions.in/about-us/ https://interactive.aljazeera.com/aje/shorts/india-menstruation-man/

38 IR 1: Steam hammer, By Unknown author (Popular Science Monthly Volume 38) [Public domain], via Wikimedia Commons, https://commons.wikimedia.org/wiki/File:PSM_V38_D349_The_nasmyth_steam_hammer.jpg

38 TR 1: The Project Gutenberg Etext of James Nasmyth's Autobiography, http://www.gutenberg.org/ebooks/476

STEP 39

39.1   Listing of Engineering Student Design Competitions, http://www.discovery-press.com/discovery-press/studyengr/competitions.pdf

39.2   "Nobel Prize facts". www.nobelprize.org. Retrieved 25 October 2018.

39.3   MLA style: Alfred Nobel – His Life and Work. NobelPrize.org. Nobel Media AB 2018. Sun. 25 Nov 2018. <https://www.nobelprize.org/alfred-nobel/alfred-nobel-his-life-and-work/>

39 IR 1: Electric Lamp adapted from: By William J. Hammer [Public domain], via Wikimedia Commons https://commons.wikimedia.org/wiki/File:Edison_incandescent_lights.jpg

39 IR 2: See page for author [CC BY 4.0 (https://creativecommons.org/licenses/by/4.0)], via Wikimedia Commons https://commons.wikimedia.org/wiki/File:Early_types_of_Electric_lightning._Wellcome_M0015309.jpg

39 TR 1: Swan K. R. Sir Joseph Swan and the Invention of the incandescent electric lamp. London: Longmans, Green and Co., 1946 pp. 21–25

STEP 40

40.1   University of Illinois, The Start-up Handbook, 2014, https://otm.illinois.edu/sites/default/files/Start-Up%20Handbook%20for%20web.pdf

40.2   Office Of Technology Development, Startup Guide, Harvard University, Https://Otd.Harvard.Edu/Upload/Files/OTD_Startup_Guide.Pdf

40.3   Definition of 3 Fs, http://lexicon.ft.com/Term?term=3-Fs

40.4   The Project Gutenberg EBook of Zeppelin, by Harry Vissering, Zeppelin-The Story of a Great Achievement,, http://www.gutenberg.org/files/32570/32570-h/32570-h.htm

40 IR 1: Rover Bicycle: By Karen Roe from Bury St Edmunds, Suffolk, UK, United Kingdom (British Motor Museum 09-2016) [CC BY 2.0 (https://creativecommons.org/licenses/by/2.0)], via Wikimedia Commons https://commons.wikimedia.org/wiki/File:1886_Starley_%27Rover%27_Safety_Cycle_British_Motor_Museum_09-2016_(29928044262).jpg

40 IR 2: Bicycle adapted from : By Agnieszka Kwiecień (Nova) [GFDL (http://www.gnu.org/copyleft/fdl.html) or CC BY-SA 3.0 (https://creativecommons.org/licenses/by-sa/3.0)], from Wikimedia Commons https://commons.wikimedia.org/wiki/File:Ordinary_bicycle01.jpg

40 TR 1: Herlihy, David V. (2004). Bicycle, The History. Yale University Press. pp. 155–250. ISBN 0-300-10418-9.

STEP 41

41.1   Darrell Zahorsky, The 7 Stages of Starting and Running a Business, November 10, 2018, The Balance Small Business, https://www.thebalancesmb.com/find-your-business-life-cycle-2951237

41.2   Agriculture and Forestry , Elements of a Business Plan, Government of Alberta, https://www.agric.gov.ab.ca/

41.3   India Abroad Person Of The Year 2013, http://im.rediff.com/news/2014/jul/09_iapoy_arogyaswami_j_paulraj.pdf

41.4   Arogyaswami Paulraj, https://web.stanford.edu/~apaulraj/

41 IR 1: Ballpoint pen adapted from: Unknown author [Public domain], via Wikimedia Commons https://commons.wikimedia.org/wiki/File:US392046-0.png

41 TR 1: US392046, US Patent,https://patents.google.com/patent/US392046

STEP 42

42.1   Elizabeth Hoyt, September 24, 2015, 20 of the Coolest College Start-ups Ever https://www.fastweb.com/student-life/articles/the-20-of-the-coolest-college-start-ups-ever

42.2   Global Startup Ecosystem Report 2018- Succeeding in the New Era of Technology, 2018, Startup Genome LLC (www.startupgenome.com).

42.3   Startup Ecosystem Rankings Report 2018, StartupBlink, https://startupservices.startupblink.com/ecosystem-report/

42.4   Kevin Sandlin, 8 Personality Traits of Every Successful Startup Founder, http://projectentrepreneur.org/inspiration/8-personality-traits-of-every-successful-startup-founder/

42.5   John Rampton, 12 Personality Traits of People Whose Startups Actually Make Money, https://www.inc.com/john-rampton/12-personality-traits-of-people-whose-startups-actually-turn-profit.html

42.6   Eric Gordon, September 1st, 2017, 5 Must-Have Traits of Startup Founders, https://www.ideatovalue.com/inno/eric-gordon/2017/09/5-must-traits-startup-founders/

42.7   Susie Poppick, Seven key traits of successful start-up founders, 23 March 2016, https://www.cnbc.com/2016/03/11/seven-key-traits-of-successful-start-up-founders.html

42.8   NASSCOM-Zinnov, "Indian Startup Ecosystem–Traversing the Maturity Cycle –2017", 2017.

42.9   Startup India Notifications,
         https://www.startupindia.gov.in/content/sih/en/startupgov/notification.ht
         ml

42.10   Personalities,
         https://archive.is/20130411215410/http://tamilnadu.com/personalities/g-d-
         naidu.html

42.11   Our Inspiration, http://www.gplast.com/our-inspiration.php

42 IR 1: Wind turbine adapted from: By Andy Dingley (scanner) [Public domain],
         via Wikimedia Commons
         https://commons.wikimedia.org/wiki/File:Blyth%27s_windmill_(Rankin_K
         ennedy,_Modern_Engines,_Vol_I).jpg

42 TR 1: Price, Trevor J. (2004). "Blyth, James (1839–1906)". Oxford Dictionary of
         National Biography
         http://www.oxforddnb.com/view/10.1093/ref:odnb/9780198614128.001.000
         1/odnb-9780198614128-e-100957

STEP 43

43.1   Jolly M, Fletcher AC, Bourne PE (2012) Ten Simple Rules to Protect Your
         Intellectual Property. PLoS Computational Biology, 8(11): e1002766.
         doi:10.1371/journal.pcbi.1002766

43.2   Drew Hendricks, 7 Simple Ways You Can Protect Your Idea From Theft,
         https://www.forbes.com/, Nov 18, 2013, 10:49am

43.3   Otis Elevator Company, About Us,
         http://www.otis.com/en/us/about/innovators/

43.4   Elisha Graves Otis, National Inventors Hall of Fame,
         https://www.invent.org/inductees/elisha-graves-otis

43 IR 1: Army tank adapted from: Ernest Brooks [Public domain or Public
         domain], via Wikimedia Commons
         https://commons.wikimedia.org/wiki/File:British_Mark_I_male_tank_Som
         me_25_September_1916.jpg

43 TR 1: Project Gutenberg's Tanks in the Great War 1914-1918, by J. F. C. Fuller

STEP 44

44.1   Rodney Katz, Michael Lewis, Prototype Design and Manufacturing
         Manual, Rev. 0, Department Of Mechanical Engineering, University of
         Victoria.

44.2   Beaudouin-Lafon, Michel & Mackay, Wendy. (2009). Prototyping Tools
         and Techniques. 1006-1031. in The human-computer interaction handbook:
         fundamentals, evolving technologies and emerging applications, L.
         Erlbaum Associates Inc. Hillsdale, NJ, USA, 2003.

44.3   User Guide, Lion Circuits, https://www.lioncircuits.com/PCB-User-Guides.html

44.4   Valdis Berzins, Software prototyping, Published in: Encyclopedia of Computer Science, 4th, Pages 1636-1638, John Wiley and Sons Ltd. Chichester, UK.

44.5   Creative Bloq, The 8 best prototyping tools for 2018, https://www.creativebloq.com/advice/the-8-best-prototyping-tools-for-2018

44.6   Bill Gates, https://www.britannica.com/biography/Bill-Gates

44.7   Profile of Bill Gates, https://www.forbes.com/profile/bill-gates/#73f8a47e689f

44.8   Gates, Bill (1996). The Road Ahead. Penguin Books. ISBN 0-14-026040-4.

44 IR 1: Nuclear fission adapted from: By Stefan- CC-BY-SA-3.0 (http://creativecommons.org/licenses/by-sa/3.0/)], from Wikimedia Commons https://upload.wikimedia.org/wikipedia/commons/5/5c/Kernspaltung.png

44 TR 1: "The Nobel Prize in Chemistry 1944". Nobel Foundation. https://www.nobelprize.org/prizes/chemistry/1944/summary

STEP 45

45.1   Master Journal List, http://mjl.clarivate.com/

45.2   SCOPUS, https://www.scopus.com/sources

45.3   Brennecke, Patricia. "Academic Integrity at MIT." Academic Integrity at MIT. Ed. Anna Babbi Klein. MIT, 2018. https://integrity.mit.edu/handbook/print-handbook

45.4   Author and reviewer tutorials, Springer, https://www.springer.com/gp/authors-editors/authorandreviewertutorials

45.5   Hays, Constance L. (June 4, 1990). "An Inventor of the Microchip, Robert N. Noyce, Dies at 62". The New York Times. https://www.nytimes.com/1990/06/04/obituaries/an-inventor-of-the-microchip-robert-n-noyce-dies-at-62.html?sec=&spon=&pagewanted=all

45.6   Leslie Berlin (2005). The Man Behind The Microchip: Robert Noyce And The Invention Of Silicon Valley. Oxford University Press. p. 235. ISBN 9780195163438.

45 IR 1: hovercraft adapted from: By Mypix [CC BY-SA 4.0 (https://creativecommons.org/licenses/by-sa/4.0)], from Wikimedia Commons https://commons.wikimedia.org/wiki/File:Hovercraft_at_Ryde,_Isle_of_Wight,_UK.jpg

45 TR 1: Patent - Bibliographic data: GB854211 (A) — 1960-11-16,
    https://worldwide.espacenet.com/publicationDetails/biblio?CC=GB&NR=8
    54211&KC=&FT=E&locale=en_EP#

STEP 46

46.1   Uddhab Bharali, the man from Assam with 118 incredible inventions,
        http://achhikhabre.com/uddhab-bharali/

46.2   This College Dropout from Assam Has over 140 Agricultural Innovations
        to His Credit". Better India. 2017-01-02.
        https://www.thebetterindia.com/81244/uddhab-bharali-assam-
        agricultural-inventions/

46 IR 1: IBM 610 adapted from: By IBM (Publicidad del aparato, publicado en
        1957) [CC BY-SA 4.0 (https://creativecommons.org/licenses/by-sa/4.0)], via
        Wikimedia Commons
        https://commons.wikimedia.org/wiki/File:610b-soft.jpg

46 TR 1: The IBM 610 Auto-Point Computer,
        http://www.columbia.edu/cu/computinghistory/610.html

STEP 47

47.1   Chuck Salter , "Failure Doesn't Suck". Fast Company. 1 May 2007.

47.2   Our story-The accidental engineer,
        https://www.jamesdysonfoundation.com/who-we-are/our-story.html

47.3   James Dyson, Against the Odds: An Autobiography Texere, 2003

47 IR 1: Intel chip adapted from: Thomas Nguyen [CC BY-SA 4.0
        (https://creativecommons.org/licenses/by-sa/4.0)], from Wikimedia
        Commons, https://commons.wikimedia.org/wiki/File:Intel_C4004.jpg

47 TR 1: The Story of the Intel® 4004, Intel's First Microprocessor,
        https://www.intel.co.uk/content/www/uk/en/history/museum-story-of-
        intel-4004.html

STEP 48

48.1   Vijay Prashanth, Here was a Baker...,
        http://www.arvindguptatoys.com/arvindgupta/bakertribute.pdf,
        Thursday, May 3, 2007

48.2   Laurie Baker: The man we will never forget,
        http://specials.rediff.com/news/2007/apr/04sld1.htm

48.3   Harikrishnan H G, Remembering Laurie Baker, the pioneer British
        architect who made India his home, posted on 1st April 2017,
        https://yourstory.com/2017/04/remembering-laurie-baker/

48 IR 1: DynaTac motorola adapted from: By Redrum0486,) [CC BY-SA 3.0
        (https://creativecommons.org/licenses/by-sa/3.0),
        https://commons.wikimedia.org/wiki/File:DynaTAC8000X.jpg

48 TR 1: MOTOROLA DynaTAC 8000x,
    https://www.imei.info/phonedatabase/7422-motorola-dynatac-8000x/

STEP 49

49.1   Christopher Pappas, Free Educational Technology, The 5 Best Free Slideshow Presentation and Creation Tools for Teachers, September 27, 2013. https://elearningindustry.com/the-5-best-free-slideshow-presentation-and-creation-tools-for-teachers

49.2   Fergal Glynn, 20 Tools for Creating and Delivering Amazing Presentations, Oct 16, 2014, https://blog.hubspot.com/marketing/presentation-tools

49.3   Wilson Greatbatch obituary, American inventor of the first practicable heart pacemaker, https://www.theguardian.com/science/2011/sep/29/wilson-greatbatch

49.4   Wilson Greatbatch". Inventor profile. National Inventors Hall of Fame.

49 TR 1: Internet of Things Global Standards Initiative, https://www.itu.int/en/ITU-T/gsi/iot/Pages/default.aspx

STEP 50

50.1   Sullia's bridge-man gets Padma Shri, https://timesofindia.indiatimes.com/city/mangaluru/sullias-bridge-man-gets-padma-shri/articleshow/56782927.cms